CIRIA C572

London, 2002

Treated ground
engineering properties and performance

Charles J A

Watts K S

CIRIA

sharing knowledge ■ building best practice

6, Storey's Gate, Westminster, London, SW1P 3AU
TELEPHONE 020 7222 8891 FAX 020 7222 1708
EMAIL enquiries@ciria.org.uk
WEBSITE www.ciria.org.uk

Summary

Ground treatment is defined as:

> the controlled alteration of the state, nature or mass behaviour of ground materials in order to achieve an intended satisfactory response to existing or projected environmental and engineering actions.

This report focuses on the properties of treated ground, where the objective of treatment has been the improvement of the load carrying characteristics of the ground. The only ground treatment methods included in this report are those that are used, or are suitable for use, in the United Kingdom.

The report aims to establish the best assessments and measurements of engineering properties and performance of treated ground, and to explain how to measure and assess such properties and performance. The report presents information that will be of direct use to geotechnical specialists, but it should also interest a broader group of people involved in building and civil engineering projects requiring ground treatment.

Guidance is given on good practice in evaluating the effectiveness of treatment. The report should lead to better use of ground treatment techniques and help to improve foundation design and construction on treated ground. Where ground treatment is used, a successful outcome depends not only on technical factors but also on the use of an appropriate contractual framework within which the treatment is procured and executed.

Treated ground: engineering properties and performance

Charles J A and Watts K S

Construction Industry Research and Information Association

CIRIA C572   © CIRIA 2002   RP604   ISBN 0 86017 572 3

Keywords		
Ground engineering, ground improvement, in-situ testing and instrumentation.		

Reader interest	Classification	
Principally geotechnical engineers, structural engineers, house builders and developers, but also those involved in building and civil projects, on which ground treatment may be required	AVAILABILITY CONTENT STATUS USE	Unrestricted Advice/guidance Committee-guided Geotechnical engineering structural and engineers, house builders, developers.

Published by CIRIA, 6 Storey's Gate, Westminster, London SW1P 3AU.

Acknowledgements

This report is the result of a research project carried out under contract to CIRIA by the Building Research Establishment Ltd. Following CIRIA's usual practice, the research project was guided by a steering group, which comprised:

Dr D H Beasley (chairman)	Halcrow Ltd
Mr T J P Chapman	Ove Arup and Partners
Mr G Fordyce	National House Building Council
Dr A Haimoni	AMEC Piling
Mr W M Kilkenny	W S Atkins Consultants Ltd (DETR Representative)
Mr C A Raison	Chris Raison Associates
Dr J M Reid	Transport Research Laboratory
Mr T Schofield	Stent Foundations Ltd
Professor N E Simons	Consultant
Mr M C Stevenson	Gibb Ltd
Dr J Wilson	W S Atkins Consultants Ltd (DETR Representative)

The CIRIA research managers for the project were Dr M Holloway-Strong and Mr F M Jardine.

The project received funding through the Partners in Innovation Scheme of the Department of the Environment, Transport and the Regions (DETR). Mr W M Kilkenny was the DETR representative on the steering group until March 2000 when Dr J Wilson became the representative.

The report was written by Dr J A Charles and Mr K S Watts, both of the Building Research Establishment Ltd. The Building Research Establishment Ltd also provided funding.

Note

Recent Government reorganisation has meant that DETR responsibilities have been moved variously to the Department of Trade and Industry (DTI), the Department for the Environment, Food and Rural Affairs (DEFRA), and the Department for Transport, Local Government and the Regions (DTLR). References made to the DETR in this publication should be read in this context. For clarification, readers should contact the Department of Trade and Industry.

CIRIA and the Building Research Establishment Ltd are grateful for the help given to this project by the funders, by the members of the steering group and by the individuals and organisations listed below who participated in the consultation processes or provided technical material:

Mr D A Baker	Balfour Beatty Major Projects
Mr N Beavor	Harris and Sutherland
Mr A P Butcher	Building Research Establishment Ltd
Mr S Cook	Nicholls Colton and Partners
Professor M C R Davies	University of Dundee
Mr N Fraser	W S Atkins Consultants Ltd
Dr D A Greenwood	Geotechnical Consultant
Dr D B Jones	Halcrow Ltd
Mr J D Maddison	Halcrow Ltd
Mr A S O'Brien	Mott MacDonald Group
Mr J J M Powell	Building Research Establishment Ltd
Dr I Robertson	Hyder Consulting
Dr D Russell	Mott MacDonald Group
Dr I Statham	Ove Arup and Partners
Mr G H Thomson	Geotechnical Consultant
Mr L Threadgold	Geotechnics
Mr R D Tinsley	Campion and Partners

CIRIA and the author[s] gratefully acknowledge the support of these funding organisations and the technical help and advice provided by the members of the steering group. Contributions do not imply that individual funders necessarily endorse all views expressed in published outputs.

Contents

Boxes 8

Figures 9

Tables 12

Glossary 11

Abbreviations 16

Notation 17

1   INTRODUCTION 21
      1.1   Background to the report 21
      1.2   Objectives of the report 21
      1.3   Scope of the report 22
      1.4   Structure of the report 22

2   REQUIRED GROUND BEHAVIOUR 25
      2.1   Types of development 25
      2.2   Soil-structure interaction 25
      2.3   Ground movements caused by foundation loading 27
      2.4   Ground movements due to causes other than foundation loading 27
      2.5   Types of structural deformation 28
      2.6   Acceptable ground deformation 30

3   REMEDYING DEFICIENCIES IN GROUND BEHAVIOUR 35
      3.1   Project responsibilities 35
      3.2   Site investigation 36
      3.3   Types of deficiency 38
            3.3.1   Inadequate strength and stiffness 39
            3.3.2   Collapse compression on wetting 39
            3.3.3   Liquefaction 40
            3.3.4   Chemical instability 42
      3.4   Management options 43
            3.4.1   Applicability of ground treatment 43
            3.4.2   Remedial processes 44

4   ENGINEERING PROPERTIES 47
      4.1   Ground properties and foundation design 47
      4.2   Index and classification properties 48
            4.2.1 Moisture content and degree of saturation 48
            4.2.2 Particle size distribution 49
            4.2.3 Plasticity indices 49
            4.2.4 Compactness 50
      4.3   Shear strength 51
      4.4   Compressibility 51

	4.5	Permeability and rate of consolidation	53
	4.6	Creep and secondary compression	56
5		**MEASUREMENT OF ENGINEERING PROPERTIES**	**59**
	5.1	Objectives	59
		5.1.1 Prior to ground treatment	59
		5.1.2 During ground treatment	60
		5.1.3 Immediately after completion of ground treatment	60
		5.1.4 In the long term	60
	5.2	Laboratory tests	61
		5.2.1 Index and classification tests	61
		5.2.2 Strength and deformation tests	62
		5.2.3 Special properties	63
	5.3	In situ tests	64
		5.3.1 Cone penetration test	65
		5.3.2 Standard Penetration Test	67
		5.3.3 Dynamic probing	68
		5.3.4 Pressuremeter test	69
		5.3.5 Flat dilatometer test	69
		5.3.6 Field vane test	69
	5.4	Geophysical tests	70
		5.4.1 Seismic methods	70
		5.4.2 Spectral analysis of surface waves (SASW)	71
		5.4.3 Continuous surface wave method	72
		5.4.4 WAK test	72
	5.5	Load tests	72
	5.6	Monitoring	74
6		**PROVISION OF GROUND TREATMENT**	**77**
	6.1	Types of treatment	77
	6.2	Role of ground treatment	79
	6.3	Basis of design	79
	6.4	Specification and site control	80
		6.4.1 Method specification	80
		6.4.2 End product specification	81
		6.4.3 Performance specification	81
	6.5	Environmental effects	81
		6.5.1 During treatment	81
		6.5.2 Subsequent to treatment	82
	6.6	Unsatisfactory performance	82
7		**IMPROVEMENT BY COMPACTION**	**85**
	7.1	Applicability of compaction methods	85
	7.2	Vibro-compaction	86
	7.3	Dynamic compaction	90
	7.4	Rapid impact compaction	96
	7.5	Compaction in thin layers	97

8   IMPROVEMENT BY CONSOLIDATION **101**
    8.1   Applicability of consolidation methods 101
    8.2   Pre-loading without installing drains 102
    8.3   Pre-loading with installation of drains 106
    8.4   Ground water table lowering 109
    8.5   Vacuum pre-loading 109
    8.6   Electro-osmosis 111

9   IMPROVEMENT BY STIFFENING COLUMNS **113**
    9.1   Applicability of stiffening columns 113
    9.2   Vibro stone columns 113
    9.3   Dynamic replacement 119
    9.4   Stabilised soil columns 120
    9.5   Vibro concrete columns 124

**10   CURRENT CAPABILITIES AND RECOMMENDATIONS FOR
     GOOD PRACTICE** **127**
    10.1 Current practice 127
    10.2 Prediction of performance 128
    10.3 Recommendations for good practice 129
    10.4 Recommendations for research 130

APPENDICES CASE HISTORIES **131**
    A.1   Refuse fill -dynamic compaction 134
    A.2   Clay fill – pre-loading 137
    A.3   Low permeability natural soil - pre-loading 140
    A.4   Mixed clay fill – vibro stone columns 143
    A.5   Miscellaneous fill – vibro stone columns 146
    A.6   Highly compressible soils - vibro concrete columns 150

REFERENCES **153**

Boxes

Box 5.1   Investigation of liquefaction resistance of PFA 63

Box 6.1   Use of vibro stone columns and vibro concrete columns for road
embankment 77

Box 6.2   Case history of factory at Warrington 83

Box 6.3   Problems at housing development in West Midlands 84

Box 6.4   Unsatisfactory performance of petrol station near Peterborough 84

Box 7.1   Vibro-compaction for housing development 87

Box 7.2   Vibro-compaction for bridge foundations 88

Box 7.3   Depth of effectiveness of dynamic compaction of clay fill 92

Box 7.4   Long-term settlement following dynamic compaction of clay fill 93

Box 7.5   Long-term settlement following dynamic compaction of old domestic
refuse 94

Box 7.6   Housing development on engineered clay fill 99

Box 8.1   Polystyrene embankment for railway use 105

Box 8.2   Effectiveness of vertical drains in consolidation of organic soil 108

Box 9.1   Vibro stone columns for wharf structure 118

Box 9.2   Vibro stone columns for road interchange 118

Box 9.3   Dynamic replacement of soft alluvial soil 120

Box 9.4   Dynamic replacement of dredged reclamation 120

Box 9.5   Lime columns in soft clay 123

Box 9.6   Lime columns in hydraulic fill 123

Box 9.7   Stabilised soil columns in silty clay with peat layers 123

Box 9.8   Vibro concrete columns with load transfer platform 126

Figures

Figure 1.1   Relationship of different sections of the report 23

Figure 2.1   Types of settlement of a tank 29

Figure 2.2   Deflection ratio in sagging and hogging 29

Figure 2.3   Building deformation 30

Figure 2.4   Criteria for onset of visible cracking of rectangular beams 32

Figure 2.5   Deflection of beams due to bending and shear 33

Figure 3.1   Flowchart for geotechnical issues 35

Figure 3.2.   Rapid impact compaction of old ash fill 37

Figure 3.3   Collapse potential of colliery spoil on submergence 40

Figure 4.1   Long-term settlement of 73 m high rock fill embankment 56

Figure 4.2   Example of the variability of domestic refuse 57

Figure 5.1   Liquefaction resistance of PFA 63

Figure 5.2   CPT truck 65

Figure 5.3   CPT in hydraulically placed PFA 65

Figure 5.4   CPT in sand with peat layer 66

Figure 5.5   Correlation of CPT-SPT ratio with particle size 67

Figure 5.6   Dynamic probing in miscellaneous fill 68

Figure 5.7   Improvement of old ash fill 68

Figure 5.8   Pad loading test 73

Figure 5.9   Vertical stress beneath loaded areas 74

Figure 6.1   Settlement of factory on vibro stone columns 83

Figure 7.1   Range of particle sizes for which vibro-compaction should be
effective 87

Figure 7.2   Monitoring settlement of houses built on vibro stone columns 88

Figure 7.3   Settlement of houses on sand with peat layer 88

Figure 7.4   Vibro-compaction of sand fill 89

Figure 7.5   Dynamic compaction of coarse fill 90

Figure 7.6   Dynamic compaction of three sand fills 92

Figure 7.7   Dynamic compaction of clay fill 93

Figure 7.8   Long-term settlement of clay fill treated by dynamic compaction . . 93

Figure 7.9   Settlement of dynamically compacted old refuse under embankment 94

Figure 7.10   Rapid impact compaction of granular fill 97

Figure 7.11   Compaction in layers of an opencast mining backfill 98

Figure 7.12   Ground treatment for distribution centre 99

Figure 8.1   Pre-loading with surcharge of fill 102

Figure 8.2   Pre-loading of clay fill 103

Figure 8.3   Settlement of houses on clay fill 104

Figure 8.4   Depth of influence of surcharge loading of fills 106

Figure 8.5   Consolidation of soft clay using vertical drains 107

Figure 8.6   Consolidation of organic soil using vertical drains 108

Figure 8.7   Vacuum pre-loading 110

Figure 9.1   Installation of vibro stone columns using the dry top-feed method . 115

Figure 9.2   Stiffening effect of stone columns 118

Figure 9.3   Installing stabilised columns 122

Figure 9.4   Installing vibro concrete columns 125

Figure A.1.1   Control testing results at Cwmbran 135

Figure A.1.2   Long-term settlement of units at Cwmbran 136

Figure A.2.1   Placing a 9 m high surcharge 138

Figure A.2.2   Settlement induced at depth by pre-loading at Corby 139

Figure A.3.1   Geological cross-section 141

Figure A.3.2   Observed and predicted settlements 142

Figure A.4.1   Skip load tests 144

Figure A.4.2   Layout of vibro stone columns at Abingdon 145

Figure A.4.3   Settlement of the building at Abingdon 145

Figure A.5.1   Fill profiles along the line of the trial foundations 147

Figure A.5.2   Dynamic probing around vibro stone columns 149

Figure A.5.3   Settlement along the strip foundations 149

Figure A.6.1   VCC load test results 151

Figure A.6.2   VCC zone test 152

Figure A.6.3   VCC zone test results 152

Tables

Table 2.1   Differential settlement limits for flat-bottomed, vertical, cylindrical storage tanks 34

Table 3.1   Collapse compression measured in non-engineered fills 39

Table 3.2   Japanese guidelines for soils requiring liquefaction evaluation 41

Table 3.3   Applicability of ground improvement for different structures and soil types 43

Table 4.1   Classification of compressibility of clays 53

Table 4.2   Typical values of constrained modulus (D) appropriate to foundation loading 53

Table 4.3   Classification of soils according to their permeability 54

Table 4.4   Typical ranges of permeability for natural soils 54

Table 4.5   Typical ranges of permeability for fills 55

Table 4.6   Properties of refuse fills 58

Table 5.1   Correlation between density index and SPT normalised blow count . . . 67

Table 6.1   Ground treatment techniques: applicability and testing methods 78

Table 6.2   Observational method applied to vibro-compaction 80

Table 7.1   Density index of sand fill deduced from in situ tests 91

Table 7.2   Field compaction of fills 95

Table 8.1   Field studies of pre-loaded fills 105

Table A.1   Summary of case histories 132

Table A.2.1   Average settlements induced by ground treatment 139

Table A.2.2   Deflection ratios for houses on different areas 139

Table A.3.1   Summary of engineering properties 141

Table A.3.2   Properties derived from pre-loading and excavation trials 143

Table A.5.1   Summary of engineering properties 147

Table A.6.1   Summary of engineering properties 151

Glossary

allowable bearing pressure
The maximum allowable net loading intensity at the base of the foundation, taking into account the ultimate bearing capacity and required margin against failure, the amount and kind of settlement expected, and the ability of the structure to accommodate this settlement.

angular distortion
The ratio of the differential settlement (Δs) between two points and the distance (L) between them.

band drain
A type of prefabricated vertical drain usually consisting of a plastic core surrounded by a geotextile sleeve; the core provides a flow path along the drain and supports the sleeve, which in turn acts as a filter separating the core and its flow channels from the soil.

bio-consolidation
Reduction in volume of refuse fills due to biodegradation.

brownfield land
Land that has been previously developed, including derelict land.

coarse soils
Gravels and sands are coarse soils; the term is applied to soils with more than about 65 per cent of sand and gravel sizes.

collapse compression
In this phenomenon a partially saturated soil undergoes a reduction in volume that is attributable to an increase in moisture content without there necessarily being any increase in applied stress.

compaction
The process of densifying soils by some mechanical means such as rolling, ramming or vibration to reduce the volume of voids.

cone penetrometer
This *in situ* testing device comprises a cone, a friction sleeve, any other sensors and measuring systems, as well as connections to push rods.

cone pressuremeter
This *in situ* testing device consists of a pressuremeter module mounted behind a cone penetrometer.

consolidation
The process of densifying soils by increasing the effective stress, using some form of static loading; consolidation of a saturated clay soil is a time-dependent process, which results from the slow expulsion of water from the soil pores.

constrained modulus
The ratio of vertical stress to vertical strain in confined compression.

creep compression

Compression that occurs under constant effective stress.

cross-hole tomography

This common form of seismic velocity tomography is carried out from two boreholes.

deflection ratio

The maximum vertical displacement (Δ) relative to the straight line connecting two points, divided by the length between the two points (L).

density index

The degree of packing in coarse soils, such as sand and gravel, can be described by the density index, I_D, which relates the *in situ* dry density of a granular fill to the limiting conditions of maximum dry density and minimum dry density; sometimes known as relative density.

fill

Ground that has been formed by material deposited through human activity rather than geological processes; it is sometimes referred to as made ground.

fine soils

Clays and silts are fine soils; the term is applied to soils with more than approximately 35 per cent of silt and clay sizes.

foundation

The part of a structure that is designed and constructed to be in direct contact with and transmitting loads to the ground.

geogrid

A type of geosynthetic with a planar structure, formed by a network of tensile elements with apertures of sufficient size to allow interlocking with the surrounding ground.

geosynthetic

A generic term for civil engineering materials such as geotextiles, geogrids, geomembranes and geocomposites that are used to modify or improve ground behaviour.

ground treatment

The controlled alteration of the state, nature or mass behaviour of ground materials in order to achieve an intended satisfactory response to existing or projected environmental and engineering actions.

hazard

A situation that, in certain circumstances, could lead to harm to the human population, the built environment or the natural environment.

heave

Upward displacement of the ground.

hogging

The mode of deformation of a foundation or beam undergoing upward bending – the opposite of sagging.

kentledge

A form of dead-weight loading, providing a reaction over a jack or directly loading a plate in a large-scale load test; it may be concrete blocks, scrap metal, containers filled with sand or water, or any other convenient material.

liquefaction
In this phenomenon a saturated sandy soil loses shear strength due to an increase in pore pressure.

low-rise buildings
Buildings not more than three storeys in height.

magnet extensometer
A device that can be installed in a borehole to monitor settlement at depth; it consists of a series of ring magnets anchored to the borehole wall and connected by plastic tubing; the position of the magnets can be located using a reed switch sensor.

Observational Method
In ground treatment, the observational method is a managed and integrated process of design, treatment control, monitoring and review that enables previously defined modifications to be incorporated during or after treatment as appropriate; the aim is to achieve greater economy without compromising technical adequacy.

organic content
The average organic matter content of a soil expressed as a percentage of the original dry weight.

pad foundation
A foundation usually provided to support a structural column consisting of a simple circular, square or rectangular slab, generally less than 6 m.

particle density
The mass per unit volume of the solid particles of a soil; sometimes termed specific gravity.

penetration test
An *in situ* test in which a device is pushed or driven into the ground while the resistance of the soil to penetration is recorded (eg standard penetration test, cone penetration test).

pre-loading
A temporary loading applied at a construction site to improve the load carrying characteristics of the ground prior to construction on the site.

primary consolidation
The reduction in volume of a fine soil caused by the expulsion of water from the soil pores and transfer of load from the excess pore water pressure to the soil particles.

raft foundation
A foundation continuous in two directions, usually covering an area equal to or greater than the base area of the structure.

relative compaction
The ratio of the *in situ* dry density to the maximum dry density achieved with a specified degree of compacton in a standard laboratory compaction test.

relative density
See density index.

resistivity
The resistivity of the soil is the electrical resistance of an element of soil, of unit cross-sectional area and unit length; its value indicates the relative capability of the soil to carry electric currents.

risk
The likelihood that a particular adverse event occurs during a specified period of time.

sagging
The mode of deformation of a foundation or beam undergoing downward bending – the opposite of hogging.

secondary compression
The reduction in volume of a fine soil caused by the adjustment of the soil structure after primary consolidation has been completed.

seismic velocity tomography
The spatial variation of seismic wave velocity is determined from external measurements of the time taken for seismic energy to travel from source to receiver.

seismic wave
Seismic waves transmit energy by the vibration of soil particles; they can be generated by some form of impact or vibration source.

serviceability limit states
These states correspond to conditions beyond which specified service criteria for a structure or structural element are no longer met.

settlement
Downward movement of a structure and its foundation resulting from movement of the ground below it; total settlement of a structure may interfere with some aspect of its functions, such as connections to services, but it is differential settlement that causes structural damage.

specific gravity
See particle density.

strip foundation
A foundation normally provided for a load-bearing wall.

tilt
Rigid body rotation of whole structure.

ultimate bearing capacity
The value of the gross loading intensity for a particular foundation at which the resistance of the soil to displacement of the foundation is fully mobilised.

ultimate limit states
These states are associated with collapse or other similar forms of structural failure.

uniformly graded soil
The soil has a majority of soil grains which are very nearly the same size.

vibro
A generic term for deep vibratory ground treatment achieved by penetration into the ground with a large vibrating poker.

well graded soil
The soil has a wide and even distribution of particle sizes.

Abbreviations

AGS	Association of Geotechnical and Geoenvironmental Specialists
AOD	Above Ordnance Datum
ASCE	American Society of Civil Engineers
BRE	Building Research Establishment Ltd
BS	British Standard
BSI	British Standards Institution
CEN	European Committee for Standardization
CLASP	Consortium of Local Authorities Special Programme
CPT	cone penetration test
CPTU	cone penetration test with pore pressure measurement (piezocone test)
CSW	continuous surface wave method
DC	dynamic compaction
DD	Draft for Development
DETR	Department of the Environment, Transport and the Regions
DMT	flat dilatometer test
DP	dynamic probing
DPC	damp-proof course
EN	European Standard
ENV	European Pre-standard
EPS	expanded polystyrene
ICE	Institution of Civil Engineers
LNG	liquid natural gas
NHBC	National House-Building Council
PFA	pulverised fuel ash
PMT	pressuremeter test
RIC	rapid impact compaction
SASW	spectral analysis of surface waves
SPT	standard penetration test
VCC	vibro concrete column
WAK	wave activated stiffness (K) test

Notation

c'	effective cohesion intercept (kPa)
c_u	undrained shear strength (kPa)
c_v	coefficient of consolidation (m^2/year)
d	length of drainage path (mm or m)
e	voids ratio
f_s	sleeve friction in cone penetration test (MPa)
h	height of sample (mm)
i	hydraulic gradient
k	coefficient of permeability (hydraulic conductivity) (m/s)
k	coefficient in equation for depth of effectiveness of dynamic compaction
m_v	coefficient of compressibility (m^2/MN)
n	porosity
n	number of impacts at a treatment point
p_L	limit pressure in pressuremeter test (kPa)
p_0	pressure required to begin to move diaphragm in flat dilatometer test (kPa)
p_1	pressure required to move diaphragm 1 mm in flat dilatometer test (kPa)
q	bearing pressure (kPa)
q_c	cone resistance in cone penetration test (MPa)
s	settlement (mm or m)
s_r	settlement ratio
t	time
t_r	thickness of foundation raft (mm or m)
u	pore pressure (kPa)
w	moisture content
w_P	plastic limit
w_L	liquid limit
z	depth below ground surface (m)
z_e	depth of effectiveness of ground treatment (m)
z_t	depth of fill (m)

A	area
A_r	area ratio
B	breadth
C_α	coefficient of secondary compression
D	constrained modulus (MPa)
D_{max}	dynamic constrained modulus computed from compression wave velocity (MPa)
D_{10}	particle size such that 10 per cent by weight of the particles are finer (mm)
D_{50}	mean particle size (mm)
D_{60}	particle size such that 60 per cent by weight of the particles are finer (mm)
E	Young's modulus (MPa)
E_a	total energy input per unit area (tonne-m/m²)
E_r	Young's modulus for raft foundation (MPa)
E_s	Young's modulus for soil (MPa)
F_c	fines content (percentage of particles finer than 0.06 mm)
G	shear modulus (MPa)
G_{max}	dynamic shear modulus computed from shear wave velocity (MPa)
H	height (m)
I_D	density index
I_L	liquidity index
I_P	plasticity index
K_{rs}	raft-soil stiffness ratio
L	length (m)
L_i	liquidity index
N	blow count in standard penetration test
N_{60}	blow count corrected to energy efficiency of 60 per cent
$(N_1)_{60}$	blow count corrected to energy efficiency of 60 per cent and for overburden stress
Q	flow rate (m/s)
R_f	friction ratio in cone penetration test; Rf = qc/fs
S_r	degree of saturation
T_v	non-dimensional time factor
U_c	uniformity coefficient; $U_c = D_{60}/D_{10}$
V_a	volume of air voids expressed as a percentage of the total volume of the soil
V_P	compression wave velocity (m/s)
V_R	Rayleigh wave velocity (m/s)
V_S	shear wave velocity (m/s)
W	weight of tamper (tonne)
α	field creep compression rate parameter
α_b	compression rate parameter for bioconsolidation in refuse fills
α_c	compression rate parameter for physical creep in refuse fills
ε_v	vertical strain
ε_{crit}	critical tensile strain in a rectangular beam
γ	bulk unit weight (kN/m³)
γ_d	dry unit weight (kN/m³)

γ_s	bulk unit weight of surcharge fill (kN/m3)	
γ_w	unit weight of water (kN/m^3)	
ϕ'	effective angle of shearing resistance	
ϕ'_{cv}	constant volume, or critical state, effective angle of shearing resistance	
ν	Poisson's ratio	
ν_r	Poisson's ratio for raft foundation	
ν_s	Poisson's ratio for soil	
ρ	bulk density (Mg/m^3)	
ρ_d	dry density (Mg/m^3)	
ρ_{dmax}	maximum dry density (Mg/m^3)	
ρ_{dmin}	minimum dry density (Mg/m^3)	
σ	normal total stress (kPa)	
σ'	normal effective stress (kPa)	
σ_v	vertical total stress (kPa)	
σ'_v	vertical effective stress (kPa)	
τ_f	shear stress at failure (kPa)	
Δ	increment	
Δ	relative deflection – maximum vertical displacement relative to the straight line connecting two points	

1 Introduction

> The objectives, scope and structure of the report are outlined. There are three main themes: the measurement of the engineering properties of treated ground, the assessment of long-term performance of treated ground, and the implications for foundation design of the actual properties and performance of treated ground. The report aims to establish and make known the best assessments of the engineering properties and long-term performance of treated ground and to explain how to measure and assess such properties and performance.

1.1 BACKGROUND TO THE REPORT

Building developments increasingly are located on brownfield sites and other types of marginal land, which have inadequate load-carrying properties. Ground treatment is likely to be required prior to construction on such sites.

This report presents the results of a CIRIA project dealing with the behaviour of ground that has been modified by the application of a geotechnical process in order to improve its load-carrying characteristics. The report focuses on the measurement of engineering properties, long-term performance, and the implications for foundation design of the actual properties and performance of treated ground.

Sources of information include previous studies in related subjects carried out by CIRIA and by BRE, other published information, and information supplied by specialist contractors, consulting engineers and NHBC. BRE has worked extensively in the area of treatment of poor ground, and data collection has built on the existing BRE database. While most of the cases that are included in the study are located in the United Kingdom, overseas experience has been utilised where there are comparable site conditions. Only ground treatment methods that are suitable for use in the United Kingdom are included in this report.

1.2 OBJECTIVES OF THE REPORT

The report aims to establish and make known the best assessments and measurements of engineering properties and long-term performance of treated ground and to explain how to measure and assess such properties and performance. While the report presents information that should be of help to geotechnical engineers, it is intended that much of it also should be of interest to a broader group of people involved in building and civil engineering projects requiring ground treatment. While geotechnical engineers may regard much of Chapter 4, which deals with engineering properties, as elementary, it is hoped that the inclusion of such material will help a broad range of readers to develop a better understanding of the properties that ground treatment aims to improve.

Guidance is given on good practice in evaluating the effectiveness of treatment. The success of ground treatment on a particular site can be evaluated in terms of:

- the satisfactory performance of the treated ground
- the cost-effectiveness of the treatment process in relation to alternative strategies
- conformity to environmental requirements.

The report should lead to better use of ground treatment techniques and help to improve foundation design and construction on treated ground. Where ground treatment is used, a successful outcome is contingent not only on technical factors but also on the adoption of an appropriate contractual framework within which the treatment is procured and executed. It should be clear where responsibility for the design, application and quality management of the treatment lies.

1.3 SCOPE OF THE REPORT

The report comprises three main themes:

1 Measurement of engineering properties of treated ground.

2 Assessment of long-term performance of treated ground.

3 Implications for foundation design of the actual properties and performance of treated ground.

No attempt has been made in this report to present a comprehensive guide to ground treatment techniques. Information on this large subject can be found in the complementary CIRIA report C573 *A guide to ground treatment* (CIRIA, 2002) and textbooks such as Van Impe (1989), Moseley (ed) (1993) and Xanthakos *et al* (1994). The main emphasis in this report is on ground treated to improve the load carrying characteristics of the ground underlying buildings and highway structures, including earth structures.

Health and safety matters are not covered in detail in this report. Further information on this subject can be found in CIRIA Special Publication 151 (Bielby, 1997). Grouting and piling are not included in the report. Geotechnical grouting has been reviewed by Rawlings *et al*, (2000).

Ground treatment is defined in CIRIA report C573 *A guide to ground treatment* as "the controlled alteration of the state, nature or mass behaviour of ground materials in order to achieve an intended satisfactory response to existing or projected environmental and engineering actions" (see CIRIA C573, 2002). The most common applications for ground treatment in the United Kingdom are in loose miscellaneous fills associated with past industrial use of land. These brownfield sites can contain domestic and industrial wastes or fills composed of soil and rock resulting from mineral excavation. For many types of development, piling through deep fill deposits is either uneconomical or impractical and some form of *in situ* treatment of the fill is necessary. The report focuses on the improvement of load-carrying characteristics, whereas CIRIA Special Publication 78 *Building on derelict land* (Leach and Goodger, 1991) covered the wider aspects of developments on derelict land including contamination.

Another area of application is associated with soft alluvial soils, sometimes overlain by shallow fills. The development of areas such as the east Thames corridor and the Severn estuary resulted in an increase in construction on soft clay soils, which may have a high organic content. These soils may have an inadequate bearing capacity and be highly compressible; traditionally, structures on these soils have been piled. Construction in these areas of many large retail and light industrial units has prompted the development of more economical techniques. Similarly the building industry has sought more economic foundation solutions in the highly competitive housing market where traditionally built masonry structures are particularly sensitive to differential settlements, but where conventional piling techniques may be uneconomical. The treatment of these soils, with the aim to provide adequate support for the structural loading, is usually designed either to pre-consolidate or reinforce the soil; alternatively it is a means to transmit structural loads through the weak soils to a sound underlying stratum.

1.4   STRUCTURE OF THE REPORT

The inter-relationship of the various sections of the report is shown in Figure 1.1.

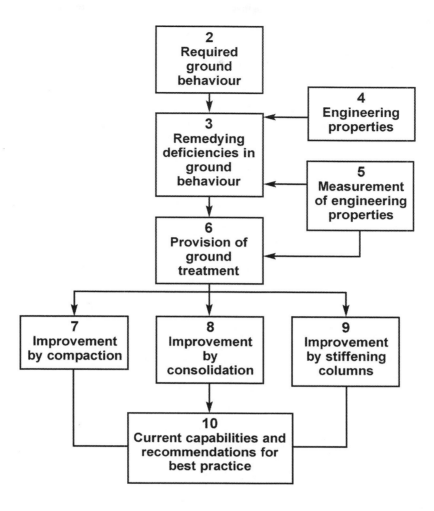

Figure 1.1   *Relationship of different sections in this report*

Ground treatment should be considered in relation to particular ground performance requirements and to ground conditions where some deficiency with regard to the proposed end use is identified. In Chapter 2, types of building and civil engineering developments are reviewed in relation to the performance requirements for the treated ground. Buildings are usually the most critical application because tolerable ground movements tend to be small and post-construction ground movements are usually the key issue. This also applies to bridge abutments and approach embankments. Ground treatment and foundation design should be closely linked.

In Chapter 3 potential deficiencies in ground behaviour are identified and their diagnosis is considered. Many treatment methods, which have been developed in different parts of the world, have been adapted for use in the wide range of fills and natural soils that are found in the United Kingdom. The treatment of compressible fills and natural soils is designed to improve the load-carrying characteristics of the ground and so minimise the total and, usually more importantly, the differential settlement of structures subsequently constructed on the treated ground. Treatment processes are categorised within a broad classification based on the underlying nature of the treatment.

Chapter 4 considers the engineering properties of treated ground that are most relevant to field performance and the appropriate ways in which these properties can be measured. Some general matters concerning ground treatment are discussed in Chapter 6. Chapters 7, 8 and 9 review the performance of treated ground through well documented case histories, which deal with improvement by compaction, improvement by consolidation and improvement by stiffening columns, respectively. The final Chapter includes an overview of current capabilities and identifies some lessons for successful ground treatment.

2 Required ground behaviour

> Developments on treated ground can include low-rise housing, light industrial and retail units with associated roads and services, high-rise buildings, large storage tanks, earth-retaining structures and embankments. The sensitivity to settlement of these various types of construction should be adequately defined to facilitate appropriate foundation design based on tolerable settlement. Throughout the report the differences between the use of ground treatment on major civil engineering works and its use on small building developments are emphasised; this section of the report has a particular emphasis on the latter use because it is where ground treatment is currently used most in the United Kingdom.

2.1 TYPES OF DEVELOPMENT

There are great differences in the scale and type of project, from small building schemes to large civil engineering structures. Developments on treated ground can include low-rise and high-rise buildings with their associated roads and services, large storage tanks, earth-retaining structures and embankments. The type of development that may be built on treated ground needs to be identified at an early stage. These various types of buildings and structures have very different deformation characteristics, two aspects being particularly significant for the selection and design of ground treatment:

1 Stiffness of the structure and its foundation (the extent to which this modifies ground deformations is considered in Section 2.2).

2 Sensitivity of the structure to ground movement (allowable settlement criteria are reviewed in Section 2.6).

Only those aspects of foundation design that are related to ground treatment are presented in this section. The general subject of foundation design and construction has been presented from a geotechnical perspective by Tomlinson (1995) and a useful manual for low-rise foundations has been written by Atkinson (1993). BS 8004: 1986, *Code of practice for Foundations* (British Standards Institution, 1986) and *Eurocode 7: Geotechnical design* (DD ENV 1997-1: 1995) (British Standards Institution, 1995) should also be consulted. Simpson and Driscoll (1998) provide a helpful commentary on Eurocode 7.

2.2 SOIL-STRUCTURE INTERACTION

Soil-structure interaction is usually complex and a pragmatic approach is to consider soil behaviour and structural behaviour in isolation from each other. However, in practice, interactive effects are experienced by all structures in contact with the ground and these interactions need to be understood.

The results of numerical analyses of rectangular rafts of different rigidities on elastic foundations were presented in graphical form by Fraser and Wardle (1976). Horikoshi and Randolph (1997) have developed the following definition of raft-soil stiffness ratio K_{rs} for a rectangular raft:

$$K_{rs} = 5.57 \frac{E_r}{E_s} \frac{(1-v_s^2)}{(1-v_r^2)} \left(\frac{B}{L}\right)^{0.5} \left(\frac{t_r}{L}\right)^3 \tag{2.1}$$

where the foundation raft has an area of $L \times B$ and a thickness t_r, E_r and E_s are the Young's moduli of the raft and soil respectively and v_r and v_s are the Poisson's ratios of the raft and soil respectively. This type of expression is helpful in assessing the degree of stiffness of a raft which is required to produce, respectively, flexible and rigid behaviour. Where the ratio K_{rs} is smaller than 0.1, behaviour is very flexible and where K_{rs} is greater than 10 it is very stiff. However, this raft-soil stiffness ratio is only relevant to the case where ground deformation is caused by the weight of the structure being transmitted through elastic foundations to ground that has linear-elastic properties.

There are some relatively simple situations where the structure and its foundation are either very stiff or very flexible:

1   When a small building is built on a stiff concrete raft, the raft may be so stiff in comparison with the soil on which the building is founded that the foundation can be considered to be fully rigid. The rigidity of the foundation should preserve the building from damaging distortion due to differential settlement and horizontal strains. The criterion for acceptable movement will relate, therefore, to the magnitude of the tilt.

2   Articulated construction has been used in areas of mining subsidence (Heathcote, 1965). The CLASP foundation solution, which depends on lightness and flexibility rather than stiffness and strength, has been used extensively (Bell, 1977). A pin-jointed steel frame structure is supported against the wind by diagonal braces incorporating compression springs in selected bays. Cladding and partitions are designed to permit the frame to deform. The foundation slab is articulated and has a plane base resting on a prepared sliding layer. Although the system was economic, its use declined because of the reduction in size of the mining industry, problems with carbonation of the precast panels and problems when buildings experienced fires (Atkinson, 1993).

3   When an embankment is built on a stiff foundation soil, the embankment may be more flexible than the ground on which it is built. The embankment deformations will follow the movement of the foundation soil and the embankments stiffness will not much influence the foundation deformation. There may be additional deformations within the embankment primarily due to compression of the fill.

In most cases the foundation will be neither fully rigid nor fully flexible and soil-structure interaction will be complex. However, full interactive analyses involving the foundation soil, the building foundation and the superstructure are so complex that they are undertaken only where there are unusual circumstances. In the routine design of simple structures the ground movements are calculated without considering interactive effects. If the calculations indicate movements that cannot be accommodated, interactive effects could be examined, but it is more likely that it will be assumed that either the ground performance needs to be improved or the structure should be modified.

In addition to the technical and analytical problems associated with soil-structure interaction, there can be professional and organisational problems. While the need for a close link between the foundation design and the ground treatment process might seem to be self-evident, it is not always present in practice. Where a building is to be constructed on treated ground, it is important that the geotechnical specialist involved with the ground treatment understands the requirements of the structural designer and that the structural designer, in turn, has realistic expectations of what can be achieved by ground treatment.

Two broad categories of ground movement need to be distinguished as the nature of soil-structure interaction is different in each case:

- ground movements caused by foundation loading
- ground movements due to causes other than foundation loading.

2.3 GROUND MOVEMENTS CAUSED BY FOUNDATION LOADING

The application of load to the ground from the weight of the structure will cause the ground to deform as the structure is built. Much routine foundation design work is concerned with the loads applied by the structure to the ground and the ability of the ground to respond to the loading without excessive ground movement.

The ultimate bearing capacity of a foundation soil is the maximum pressure that the soil is capable of carrying. The allowable bearing pressure is the maximum allowable net loading intensity taking into account not only the ultimate bearing capacity, but also the estimated amount and rate of settlement and the ability of the structure to accommodate the settlement. It is necessary to check that allowable bearing pressure is not exceeded. In many cases the ground has adequate bearing capacity and the ground deformation is related to the one-dimensional compression behaviour, which is modelled in a laboratory oedometer test as outlined in Section 5.2.2.

Ground treatment is frequently required for large floor slabs in warehouses. The variable floor loadings are not foundation loadings as such and will be applied after completion of the building when it is brought in to use. A loading of 80 kPa is often specified for some structures, but the real loads may be far less because the entire floor area is extremely unlikely to be loaded to the full 80 kPa. In some cases there are very onerous loading requirements; for example, where a loading of 50 kPa is applied uniformly over a large floor slab and the building itself is piled, significant differential settlement is likely to occur between the floor slab and the walls of the building. The serious implications of this type of loading mean that it is important that design loadings are realistic.

Most of the cases where soil is deemed to require treatment because it is too compressible will be in the high compressibility and very high compressibility soil categories as described in Section 4.4 and Table 4.1. This will include some non-engineered fills, estuarine clays, organic alluvial clays and peats. Settlement should be evaluated for realistic loading conditions with an achievable level of ground improvement in view.

The magnitude and timing of ground movements are major influences on structural performance and the incidence and severity of structural damage. The rate at which the movement takes place is of lesser significance in relation to damage, but is of importance when assessing building movements. For granular soils, movements due to foundation loading will largely take place as the structure is being built, whereas on clay soils substantial long-term settlement may continue after construction, although at a rate that decreases with time. The greatest cause for concern is associated with movement initiated subsequent to completion of construction and this is most likely to result from causes other than the weight of the structure.

2.4 GROUND MOVEMENTS DUE TO CAUSES OTHER THAN FOUNDATION LOADING

Ground movements may occur due to factors unrelated to the weight of the structure and such movements can have a crucial effect on structural performance. Some of the potential causes of post-construction ground movements are:

- increases in effective stress caused by the addition of fill
- loss of lateral support caused by adjacent excavation
- changes in effective stress caused by variations in ground water level; a lowering of the ground water level will result in an increase in effective stress and consequent settlement whereas a rising ground water level will reduce the effective stress, can cause heave and reduce allowable bearing pressures
- in poorly compacted fills, collapse compression can occur due to an increase in moisture content associated with either water penetrating downwards from the ground surface or a rising ground water level
- loss of fines due to dewatering
- decomposition of organic matter, particularly in biodegradable fills; substantial reductions in volume cause large settlements at the ground surface and remove lateral support to treatment systems such as vibro stone columns
- chemical reactions within certain types of fill, eg steel slag, pyritic shale; volume changes may result in heave of the ground surface
- dynamic loading including vibration can cause loose granular soils to liquefy and/or reduce in volume
- swelling and shrinkage due to soil moisture content changes, sometimes caused by action of trees in clay soils
- subsidence associated with mining, tunnelling and collapse of natural solution features.

2.5 TYPES OF STRUCTURAL DEFORMATION

Settlement is usually the critical factor for foundation performance although horizontal ground movements can also be significant. Acceptability criteria for ground movements may be in terms of total settlement, differential settlement, angular distortion or tilt.

Foundation movement in the form of settlement (s) is normally quantified in terms of differential settlement (Δs), which is the maximum vertical displacement of one part of the structure with respect to another. Three types of foundation movement are illustrated in Figure 2.1 for the simple case of a storage tank:

(a) uniform settlement	$s_A = sB$	$\Delta s = 0$
(b) uniform tilt	$s_A < sB$	$\Delta s = s_B - s_A$
(c) non-uniform tilt	$s_A < sB$	$\Delta s = s_B - s_A$

A structure should not be adversely affected by uniform settlement, but there could be problems with connections for services if the ground surrounding the structure does not undergo similar settlement. A small uniform tilt should not distort a building, but excessive tilt may cause structural damage and some tall structures could become unstable. In some cases, processes taking place in a building can be affected by differential settlement or tilt. Non-uniform tilt can lead to distortion and serious damage to a structure.

Figure 2.1 illustrates how differential settlement can be associated with either tilt or distortion. The maximum differential settlement might be the same in cases (b) and (c) in Figure 2.1, yet the effect on the structure could be very different as the distortion in (c) would be much more serious than the uniform tilt in (b). Distortion could occur, of course, without tilt. The distortion of a structure, therefore, cannot be directly related to differential settlement.

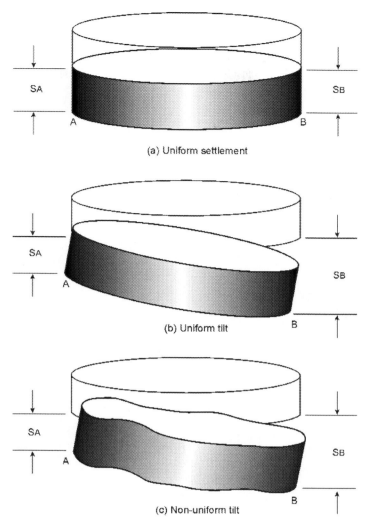

Figure 2.1 *Types of settlement of a tank (after Ishihara and Furukawazono, 1999)*

Distortion can be defined in various ways. Skempton and MacDonald (1956) used angular distortion, defined as the ratio of the differential settlement (Δs) of two points and the distance (L) between them. The most meaningful parameter in the context of cracking is deflection ratio, which is defined as the maximum vertical displacement (Δ) relative to the straight line connecting two points divided by the length (L) between the two points (Burland and Wroth, 1974), see Figure 2.2. Here, the section of the building denoted L_{AD} is undergoing sagging and the section denoted L_{DF} is undergoing hogging.

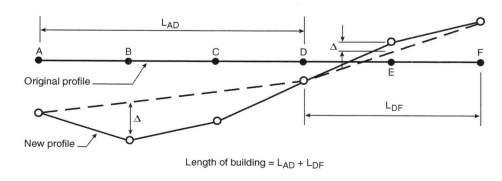

Figure 2.2 *Deflection ratio in sagging and hogging (after Burland and Wroth, 1974)*

Figure 2.3 shows the deflection ratio for the simple case of a building undergoing a sagging deformation. There is deformation of the building without any tilt occurring. With traditional brick and masonry structures, damage will be much more severe where ground deformations give rise to hogging, or upward bending, rather than the downward bending that occurs in sagging.

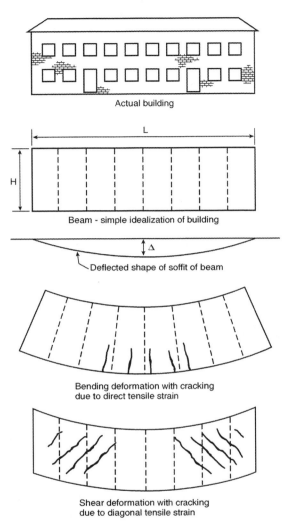

Actual building

Beam - simple idealization of building

Deflected shape of soffit of beam

Bending deformation with cracking due to direct tensile strain

Shear deformation with cracking due to diagonal tensile strain

Figure 2.3 *Building deformation (after Burland and Wroth, 1974)*

2.6 ACCEPTABLE GROUND DEFORMATION

Ground treatment should be designed to provide long-term stability for the structure and acceptable performance with respect to finishes and service connections. For most types of construction on treated ground, the durability and satisfactory long-term performance of the treated ground will be of primary importance and crucial to an assessment of whole-life costs. Failure to achieve this principal objective may be very difficult to identify in the short-term and will only be apparent in the medium to long-term.

Whether ground movements are due to the weight of the structure or to effects not associated with the structure, some movement of the structure will occur and the stiffness of the structure will modify the ground behaviour. Foundation design limit states associated with bearing capacity and settlement are relevant to structural performance. Serviceability limit states corresponding to conditions beyond which specified service criteria for a structure or structural element are no longer met should be distinguished from ultimate limit states associated with collapse or other forms of structural failure.

The required performance of the treated ground depends on the type of development. For example, where there are variable ground conditions, buildings with a large plan area are more vulnerable than small buildings with stiff foundations. Masonry construction is more susceptible to damage from differential ground movements than relatively flexible steel frame construction. The required performance of the treated ground also depends, to some extent, on the length of the working life of the structure.

The long-term performance criterion for a settlement-sensitive structure might be expressed in terms of very small limiting values for distortion and tilt. In contrast, the required long-term performance for treated ground supporting an embankment will depend on the function of the embankment and could be little more than that the stability of the embankment is not impaired. Although buildings are usually the most critical situation, embankment settlement becomes significant at structures, for example piled bridge abutments, and in relation to flood protection levels.

Factors that can affect how much foundation movement is acceptable for a building include:

- visual appearance
- serviceability (including external service connections)
- structural integrity.

In many cases allowable foundation movements are limited by their effect on visual appearance and serviceability, rather than the integrity or stability of the structure. Serviceability is often the prime consideration for industrial buildings and depends on the function of the building, the reaction of the user and owner, and economic factors such as value, insurance cover and the importance of first cost in relation to whole-life cost. In other cases, the level of tolerable movement will be dictated by a particular function of the building or one of its services (eg the proper operation of overhead cranes, lifts or precision machinery, the use of high stacking forklift trucks). For domestic buildings, visual appearance may be critical; cracking of walls, cladding materials and distortion with consequential sticking windows and doors should be avoided. The effect of ground movements on infrastructure including roads, drains and services also needs to be considered, particularly drainage falls.

The cracking of walls is dependent on a wide range of factors and, in particular:

- the length-to-height ratio of the wall (L/H)
- the mode of deformation, whether the wall is hogging or sagging.

From loading tests on brick walls carried out at the Building Research Station, Meyerhof (1953) concluded that differential settlement corresponding to $\Delta/L = 0.5 \times 10^{-3}$ would give a reasonable margin of safety. For brick dwellings, Polshin and Tokar (1957) suggested that the allowable settlement corresponded to $\Delta/L = 0.3 \times 10^{-3}$ for $L/H < 2$ and $\Delta/L = 1.0 \times 10^{-3}$ for $L/H = 8$.

Damage criteria were developed by Burland and Wroth (1974) based on a critical tensile strain, $\varepsilon_{crit} =$ of 0.075 per cent for brickwork and other materials. As expected, the hogging criterion is the most critical; for $L/H < 3$, the criterion for the onset of visible cracking of rectangular beams corresponds approximately to $\Delta/L = 0.2 \times 10^{-3}$.

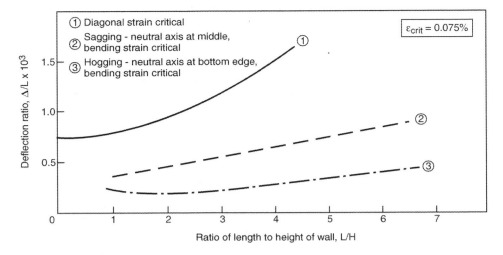

Figure 2.4 *Criteria for onset of visible cracking in rectangular beams (after Burland and Wroth, 1974)*

The results presented by Burland and Wroth (1974) in the form of a plot of Δ/L versus L/H for a particular value of ε_{crit} = 0.075 per cent were generalised by Boscardin and Cording (1989), as shown in Figure 2.5, as a plot of $\Delta/(L.\varepsilon_{crit})$ versus L/H. The importance of direct horizontal extension in initiating damage should be recognised.

In specifying acceptable deformation criteria, there is a trade-off between simplicity and relevance. Total settlement is the simplest form of movement to predict. In many circumstances, particularly on brownfield sites, it is difficult to predict total settlement let alone differential settlement.

There has been, therefore, a tendency to depend on correlations between measurements of maximum total settlement and manifest damage:

- Terzaghi and Peck (1948) stated that most ordinary structures such as office buildings and factories can withstand a differential settlement between adjacent columns of 20 mm and suggested that for footings on sand the differential settlement is unlikely to exceed 75 per cent of the maximum settlement.

- Design limits for the total settlement of isolated foundations of 40 mm on sand and 65 mm on clay were proposed by Skempton and MacDonald (1956); the smaller value for sand was associated with the perception that variability was more likely and hence differential settlement could be a greater proportion of total settlement.

- Sowers and Sowers (1970) quoted a maximum permissible total settlement of 150 mm to 600 mm for drainage and for access.

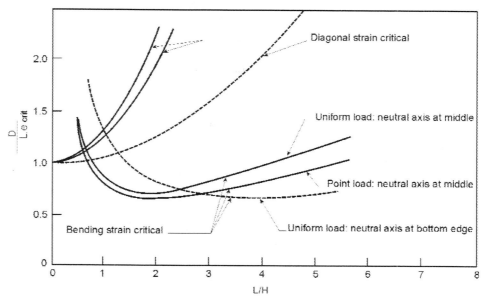

Figure 2.5 *Deflection of beams due to bending and shear (after Boscardin and Cording, 1989)*

The validity of correlations between distortion, differential settlement and total settlement is questionable. On very variable filled ground, the differential settlement may be close to the magnitude of the total settlement. Terzaghi (1956) criticised the attempt to relate building distortion to maximum settlement by Skempton and MacDonald (1956) and observed that, in his judgement, the authors' conclusions regarding the ratio between angular distortion and maximum settlement were too sweeping to be accepted.

Boone (1996) has pointed out that attempts to relate the onset of damage to a single deformation criterion exclude many important factors such as:

- flexural and shear stiffness of building sections
- nature of the ground movement profile and the location of the building within the profile
- degree of slip between the foundations and the ground
- building configuration.

Where small structures are built on treated ground, stiff foundations can be provided relatively cheaply. Where the foundations are so stiff that the structure does not deform but simply tilts and where the superstructure is not subjected to damaging tensile forces, an acceptable settlement criterion can be defined in terms of tilt. The problems caused by tilt will depend on the type of building and its purpose:

1 For low-rise buildings tilt is likely to become noticeable at around 1/200; a suitable limit for an acceptable tilt for design purposes in a typical case is 1/500 (Skinner and Charles, 1999).

2 For buildings containing some types of specialist equipment there may be very small tolerance of tilt. In such cases structural requirements need to be carefully examined because, for example, a tilt criterion of 1/2000 for a warehouse containing high racking might rule out a ground treatment solution. Lambe and Whitman (1979) quote an extreme case where a tilt of 1/50 000 could destroy the usefulness of a radar system; ground treatment is most unlikely to be adequate for such an application.

3 For cylindrical storage tanks BS 7777: Part 3: 1993 (British Standards Institution, 1993) provides the differential settlement limits reproduced here as Table 2.1 for guidance.

Table 2.1 *Differential settlement limits for flat-bottomed, vertical, cylindrical storage tanks (Table 3 from BS 7777: Part 3: 1993)*

Type of settlement	Differential settlement limit
Tilt of the tank	1:500
Tank floor settlement along a radial line from the periphery to the tank centre	1:300
Settlement around the periphery of the tank	1:500 but not exceeding the maximum settlement limit calculated for tilt of the tank

3 Remedying deficiencies in ground behaviour

> Prior to considering ground properties and different types of ground treatment, it is necessary to identify the deficiencies in load-carrying characteristics that require to be remedied. Appropriate site investigation is essential and identification of deficiencies forms part of the investigation process. The types of problem that may be encountered and the ways in which they may be identified are described. This needs to be set in the context of the project organisation. The objective of ground remediation should be to eliminate, or at least substantially mitigate, the identified deficiencies in ground behaviour. Management options and various strategies for remediation are reviewed.

3.1 PROJECT RESPONSIBILITIES

All constructed facilities come into contact with the ground and the complexity of the soil-structure interaction should influence how the project is planned, designed and built. Not all risk for subsurface conditions can be avoided or eliminated, but risks are manageable and controllable provided that they are identified in time (Hatem, 1998). Where ground conditions are poor, the risks for those involved in the development can be high and it is particularly important that ground engineering issues are properly evaluated. A flowchart showing the relationship of these issues is given in Figure 3.1.

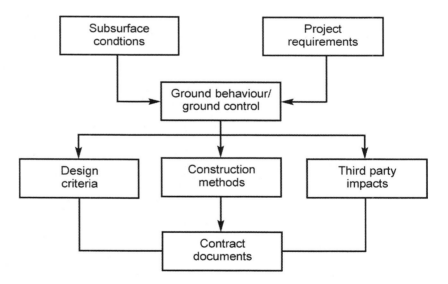

Figure 3.1 *Flowchart for geotechnical issues (after Brierley, 1998).*

The successful outcome of a project requires not only technical competence, but also an appropriate organisational structure in which the responsibilities of the various parties are identified and which facilitates effective communication between those parties. The parties involved in a ground treatment project will depend on the form of the contract as well as the type and size of the project, but will normally include the developer or promoter of the scheme, the project designer, a site investigation contractor and a ground treatment contractor. While it is not the purpose of this report to recommend appropriate contractual arrangements, there are some common elements concerning project responsibilities that need to be emphasised with respect to the provision of geotechnical expertise. CIRIA C573 (CIRIA, 2002) describes the respective roles of the promoter of the project, the ground improvement designer and the specialist contractor.

Geotechnical expertise is required for the successful application of ground treatment and the way in which the ground treatment will be provided needs to be understood by all the parties. Early involvement of a geotechnical engineer is vital. While the importance of the geotechnical specialist is unlikely to be overlooked in large-scale civil engineering projects, it may not always be fully appreciated in smaller-scale building developments. Responsibilities should be clearly defined and have a close correlation with capabilities. Effective intra and inter-team communications is important on all projects.

Construction (Design and Management) Regulations (Health and Safety Commission, 1994) apply to all stages of a construction project, including ground treatment. It is essential for safe and effective execution of treatment works that all relevant requirements are met, including provision of an appropriate risk assessment for each project as required by the Management of Health and Safety at Work Regulations (Health and Safety Commission, 1992).

3.2 SITE INVESTIGATION

The ground is a major area of risk in a construction project. Adequate site investigation and, where appropriate, ground treatment should reduce that risk. However, even where reasonable efforts are made to investigate subsurface conditions at the site, the actual ground conditions encountered during construction may be more variable than indicated by the investigation, particularly in brownfield sites. The identification of deficiencies in load-carrying properties forms one aspect of site investigation, which is particularly relevant to ground treatment.

Wider issues involved in site selection and investigation, including constraints of the natural environment, planning permission and personal indemnity insurance, are described by Lampert and Woodley [Eds] (1991). A general account of site investigation practice can be found in Clayton *et al* (1995). Authoritative information on particular matters is contained in reports and codes prepared by the British Standards Institution (BSI), European Committee for Standardisation (CEN), the Institution of Civil Engineers (ICE) and the Association of Geotechnical and Geoenvironmental Specialists (AGS):

- BS 5930:1999 *Code of practice for site investigations* (British Standards Institution, 1999)
- BS 10175:2001 *Code of practice for the Investigation of potentially contaminated sites* (British Standards Institution, 2001)
- DD ENV 1997-3:2000 Eurocode 7: *Geotechnical design – Part 3: Design assisted by field testing* (British Standards Institution, 2000b)
- *Site investigation in construction* (Institution of Civil Engineers, Site Investigation Steering Group, 1993)
- *Code of conduct for site investigation and guidelines for good practice in site investigation* (Association of Geotechnical and Geoenvironmental Specialists, 1998a, 1998b).

The site investigation should include a desk study, site reconnaissance and a ground investigation. The need for treatment will become apparent during the investigation process and a further phase of the investigation is likely to be required in relation to the chosen ground treatment. Field trials can yield reliable information about actual ground behaviour that is difficult to obtain in other ways. However, cost considerations usually restrict this type of investigation to large projects. Properly designed and carefully executed reinstatement of ground opened up in the ground investigation is important to avoid leaving local weak zones.

Site investigation for low-rise buildings presents particular problems due to the small scale of many developments and the frequent lack of geotechnical input. BRE has issued digests giving guidance on best practice for site investigation for low-rise building:

- 318 – desk studies (Building Research Establishment, 1987a)
- 322 – procurement (Building Research Establishment, 1987b)
- 348 – the walk-over survey (Building Research Establishment, 1989a)
- 381 – trial pits (Building Research Establishment, 1993a)
- 383 – soil description (Building Research Establishment, 1993b)
- 411 – direct investigations (Building Research Establishment, 1995).

Accurate and relevant site investigation information is vital for successful design and application of ground treatment. Ground conditions should be adequately characterised and this is particularly difficult with heterogeneous mixed fills. The soil properties required for the design of the treatment should be determined; Table 3.1 in CIRIA C573 describes the geotechnical information needed for different treatments.

With some forms of ground treatment the process itself forms an additional site investigation. For example, monitoring power consumption and stone consumption during the installation of vibro stone columns can give valuable additional data concerning site conditions. Figure 3.2 illustrates how rapid impact compaction at 1.75 m centres identified a diagonal feature across a 35 m × 40 m trial area of an old ash fill.

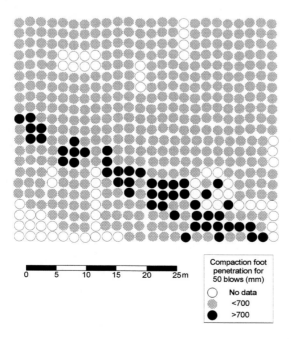

Figure 3.2. *Rapid impact compaction of old ash fill (after Watts and Charles, 1993)*

3.3   TYPES OF DEFICIENCY

The first step in any consideration of ground treatment should be the identification of some deficiency in the ground that requires treatment; the importance of correct diagnosis cannot be over-emphasised. The effectiveness of the available treatment methods can then be assessed. Inevitably the cost of alternative treatments will be considered, but only those techniques that actually can improve the identified deficiencies in the load-carrying properties of the ground should be considered. It may seem unnecessary to give such emphasis to the importance of the identification of ground problems, but failure to define the hazard adequately can lead to serious difficulties at a later stage and inadequate performance of the treated ground. Examples of this are given in Section 6.6.

An indication of deficiencies that may require treatment is given in this section; the engineering properties, which control the load-carrying characteristics of the treated ground, are described in Chapter 4. In most situations the problem with the ground relates to the potential for unacceptably large movements of a structure, usually differential settlement. Deficiencies in the ground should be evaluated in relation to the proposed type of development as described in Section 2. Deficiencies are usefully reviewed in relation to particular types of ground and much ground treatment work is carried out on fills, soft clays and organic soils.

The term *fill* is used to describe ground that has been formed by human activity rather than geological processes. A survey carried out by BRE in 1992 indicated that nearly 20 per cent of low-rise construction in Great Britain was taking place on filled ground. This percentage is likely to increase because of pressure to locate building developments on brownfield sites, many of which contain substantial depths of fill. The greatest use of ground treatment in the United Kingdom is to improve the properties of fill.

BS8004 *Code of practice for foundations* (British Standards Institution, 1986) warns that "All made ground should be treated as suspect because of the likelihood of extreme variability". The other basic deficiency of non-engineered fills is a lack of adequate compaction, which results in the fill being in a relatively loose condition. Many brownfield sites contain old structures on or in the ground, which can create hard spots and obstructions, resulting in a high potential for large differential settlements. For non-engineered fills, problems are principally associated with heterogeneity and a loose condition. Consequent deficiencies can include vulnerability to collapse compression on inundation and liquefaction under dynamic loading.

BRE Digest 427 (Building Research Establishment, 1997, 1998a and 1998b) provides guidance on fill for low-rise buildings. Virtually all the problems associated with inadequate strength or stiffness, or with collapse or liquefaction potential, will be remedied, or at least ameliorated, by densification of the fill. This is usually achieved by the compaction methods described in Chapter 7. Pre-loading (Section 8.2) is likely to be effective in partially saturated fills and high permeability saturated fills without any necessity to install vertical drains. The installation of properly constructed vibro stone columns (Section 9.2) in coarse fills will result in an increase in density around the columns.

Other common types of marginal ground that may require treatment include soft clays and organic soils. These soils are likely to have a low undrained shear strength and to be highly compressible; remedies can involve consolidation by pre-loading combined with the installation of vertical drains (Section 8.3), or reinforcement of the soil, using, for example, vibro concrete columns (Section 9.5) or stabilised soil columns (Section 9.4).

3.3.1　Inadequate stiffness and strength

The most obvious form of deficiency is an inability to support the weight of the structure, without excessive ground deformation, because the soil has high compressibility or low shear strength and bearing capacity. For example, soft clay soils may be highly compressible and also may have low undrained shear strength. This type of deficiency is considered in more detail in sections 4.3 and 4.4.

3.3.2　Collapse compression on wetting

Collapse compression is the term used to describe situations in which a partially saturated soil undergoes a reduction in volume that is attributable to an increase in moisture content. Collapse compression on wetting or inundation is a widespread phenomenon affecting both natural soils and fills, which can occur without any increase in applied total stress. Most types of partially saturated fill are susceptible to collapse compression under a wide range of applied stress when first inundated if they have been placed in an insufficiently dense and/or wet condition. If this occurs after construction on the ground, buildings can suffer serious damage. This often represents the most serious hazard for buildings on fill.

Various mechanisms may cause collapse compression. Dudley (1970) concluded that for collapse to occur the soil must start with a structure that is open, with large voids ratio for the particular materials and must have a temporary source of strength to hold the soil grains in position against shearing forces. These temporary sources of strength could be capillary tensions or cementing agents, and are reduced by addition of water.

Two major field experiments at restored opencast mining sites, which were started by BRE in the 1970s, showed that collapse compression on inundation is a major hazard:

- At Horsley in Northumberland, during a three year period, the ground water level rose 34 m through a 70 m deep mudstone and sandstone backfill; during this period the ground surface settled by 0.3 m and locally collapse compression was as large as 2 per cent (Charles *et al*, 1977, 1993).
- At Corby, water was added to the backfill via surface trenches; the upper half of the 24 m deep fill was predominantly clay fill; the ground surface settled by as much as 0.2 m and locally the compression was as large as 6 per cent (Charles *et al*, 1978; Burford and Charles, 1991).

Prior to this work, collapse compression was not generally recognised as a hazard for opencast backfills in the United Kingdom. Collapse compression measured at a number of filled sites is summarised in Table 3.1.

Table 3.1　*Collapse compression measured in non-engineered fills*

Fill type	Collapse compression	References
Mudstone/sandstone	2%	Charles et al, 1977; 1993
Clay/shale fragments	5%	Smyth-Osbourne and Mizon, 1984
Stiff clay	6%	Charles et al, 1978; Burford and Charles, 1991
Stiff clay	3%	Charles and Burford, 1987
Colliery spoil	7%	Skinner et al, 1997

Figure 3.3 shows the collapse potential determined at low stresses in laboratory oedometer tests on a colliery spoil fill. The contours of collapse compression indicate the collapse compression that will occur when the fill is submerged.

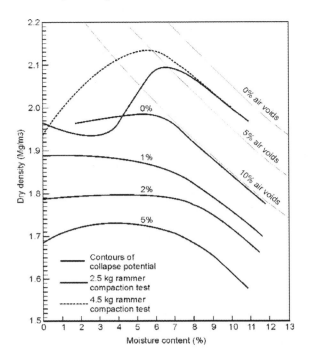

Figure 3.3   *Collapse potential of colliery spoil on submergence (after Skinner et al, 1999)*

It should be assumed that a non-engineered fill above the water table has collapse potential unless there is a good reason to believe otherwise, for example if the fill has previously been submerged. Where a soil has collapse potential, it should be recognised that inundation of the ground may well occur during the lifetime of the structure. The increase in moisture content of the fill, which triggers collapse compression, can be caused either by downward infiltration of surface water or by a rising ground water level. Settlement of fill may cause drains to fracture causing ingress of water, which can lead to much larger collapse settlements.

Reduction of collapse potential is a critical consideration in the ground treatment of many fills. If the air voids are reduced to 5 per cent or less, collapse compression is highly unlikely. The commonly adopted relative compaction criterion of 95 per cent of a maximum dry density is less certain to be adequate (Charles *et al*, 1998).

3.3.3   Liquefaction

When saturated sandy soil loses shear strength due to an increase in pore pressure with a corresponding reduction in effective stress, this is termed liquefaction. The term is sometimes restricted to a complete loss of shear strength, when the effective stress drops to zero, but it is also used to denote a partial loss of strength due to build-up of pore water pressure. Cyclic loading of saturated sands under conditions of no drainage, results in a marked reduction of strength associated with an increase in pore pressure. There appears to be a threshold strain required to produce liquefaction. Below this threshold, cyclic straining is non-destructive and there is no rearrangement of soil grains. Liquefaction can occur in both natural soils and fills. More detailed information on the physics and mechanics of soil liquefaction can be found in the proceedings of a recent international workshop (Lade and Yamamuro [Eds], 1998).

Loose, sandy soils may be subject to loss of strength and associated reduction in volume when subjected to dynamic loading, associated with one of the following:

- earthquakes
- traffic vibrations
- machine foundation vibrations
- pile driving
- blasting.

The United Kingdom is an area of low seismicity where the probability of earthquakes causing significant damage is relatively low. Prior to 1970, the only structures built in Great Britain that allowed for significant earthquake forces were a number of dams in an area of Scotland that had suffered lengthy and repetitive swarms of moderately sized earthquakes.

Following on from the work of the nuclear industry, seismic risk is being addressed increasingly in the design of major structures. A review of seismic hazard and the vulnerability of the built environment was carried out by Ove Arup and Partners (1993). A seismic zone map of the United Kingdom has been provided by Musson and Winter (1996 and 1997) with contours of peak ground acceleration for earthquakes with an annual exceedance probability of 1×10^{-4}.

Practical methods of evaluating liquefaction potential of sand deposits during earthquakes are of major concern in seismically active regions. It is difficult to determine the *in situ* density of sands in the field and therefore density index and liquefaction potential are often inferred from penetration testing. *In situ* test methods are described in Section 5.3.

Vulnerability to liquefaction during earthquakes and the measures that can be taken against liquefaction have been comprehensively reviewed by the Japanese Geotechnical Society [Eds] (1998). Table 3.2 summarises Japanese practice in defining those soil types where liquefaction evaluation is required in terms of mean particle size, D_{50}, fines content, F_c, and uniformity coefficient, U_c.

Seed (1987) considered that there were two main problems confronting the geotechnical engineer dealing with a situation where soil liquefaction may occur:

1 Determining the stress conditions required to trigger liquefaction.
2 Determining the consequences of liquefaction in terms of potential sliding and potential deformations.

Table 3.2 *Japanese guidelines for soils requiring liquefaction evaluation (after Japanese Geotechnical Society [Eds], 1998; Table 3.4)*

Application	Soil type
LNG in-ground storage	$D_{50} < 2$ mm, $F_c < 35\%$
Highway bridges	0.02 mm $< D_{50} < 2$ mm
Port and harbour facilities	Sands with low U_c: 0.02 mm $< D_{50} < 2$ mm Sands with high U_c: 0.01 mm $< D_{50} < 3$ mm
Building foundations	$F_c < 35$ per cent Silt with low plasticity Silt with moisture content close to liquid limit Gravel with significant fines content Gravel sandwiched between impermeable soils

Density index, I_D, is an important parameter of the sand in determining resistance to liquefaction. (This parameter is defined in Section 4.2.4.) The build-up in pore pressure is considered to be due to a potential reduction in volume produced by the cyclic loading. Thus one requirement for liquefaction is that densification is possible. A soil with greater I_D requires greater energy for liquefaction and it is not likely to occur in dense fills due to their tendency to dilate.

The length of time during which the excess pore pressures remain is a function of the permeability of the material. A high permeability will mean that they dissipate quickly and may not even develop. Liquefaction is thus less likely in gravelly soils due to their greater permeability. The particle size distribution for the most easily liquefiable soils is between 0.07 mm and 0.6 mm. Thus sandy materials are most susceptible, while an increased fine content reduces the tendency to liquefaction. The boundary for potentially liquefiable soils is sometimes quoted as between 0.02 mm and 2.0 mm.

Generally sandy and silty hydraulic fills have been reckoned to be susceptible to liquefaction under dynamic loading. Although this has been of concern mainly in the more seismically active regions of the world, Bell *et al* (1986) have described a site at Barrow-in-Furness where hydraulic sand fill was densified to improve liquefaction resistance to earthquakes. Humpheson *et al* (1991) carried out a series of cyclic undrained triaxial compression tests on undisturbed and resedimented samples to demonstrate that hydraulically placed PFA fill at Peterborough would not be vulnerable to liquefaction under the maximum strength earthquake predicted for the area.

If ground improvement is required prior to construction, it is sensible to select an improvement technique that is closely related to the perceived hazard. If liquefaction under dynamic loading is the major hazard, an improvement technique involving dynamic loading such as vibro-compaction or dynamic compaction would be appropriate.

3.3.4 Chemical instability

There are cases where the chemical composition of soils is of importance in civil engineering and building developments. For example, the presence of organic matter can be a problem and sulphate content may be important. Ground treatment is often used on brownfield land where there can be extensive chemical contamination. Chemical properties may be of particular interest with regard to the stability and durability of some forms of ground treatment. The material used to form vibro stone columns should be selected with care; eg limestone may not be suitable in acidic ground conditions. Some foundation problems, which are related to ground chemistry, are described in *Ground chemistry implications for construction* (Hawkins [Ed], 1997).

Where a treatment process involves the chemical modification of soils, the durability of the process needs to be established. Similarly, where physical reinforcement is to be introduced into the ground, the durability of the reinforcing elements needs to be assessed. The likelihood of the corrosion of metal inclusions can be related to chemical composition, electrical resistivity and redox potential of the soil.

In some circumstances there can be problems with chemical stability associated with densification techniques, which trigger chemical reactions in fills. Eakin and Crowther (1985) have described the reclamation of a 40 ha steelworks waste tip, which contained iron and steel slags at Brenda Road, Hartlepool. Ground treatment involved excavation and refilling under controlled conditions to form a 4 m depth of compacted fill. In the first three years following treatment, volume expansions of up to 3 per cent were recorded and a number of chemical reactions were identified, which could have caused the volumetric instability.

3.4 MANAGEMENT OPTIONS

Once the deficiencies in the load-carrying characteristics of the ground are correctly diagnosed, it should be possible to select and implement a technically adequate solution. There are a number of alternative strategies including:

- relocation of the development; the use of the site for the intended purpose is either abandoned or substantially revised by locating buildings on better areas

- removal of the poor ground; this solution involves either the physical removal of the problem soils to another site and, if necessary, placement of an imported engineered fill, or excavation, sorting to facilitate removal of unacceptable material and recompaction of remaining acceptable material as an engineered fill

- deep foundations; the problem soils are by-passed by founding in a firmer stratum at greater depth, often using piles

- ground treatment; the improvement of the engineering properties of the ground

- relaxation of structural requirements.

The selection of a remedial strategy involves many factors related to technical adequacy, cost and environmental effects. This report is restricted to ground treatment solutions.

3.4.1 Applicability of ground treatment

Ground treatment may be selected on cost grounds, but it is essential that its technical adequacy is established. It is unlikely that ground treatment will be as effective as piling in reducing settlement. The applicability of ground treatment for different ground conditions and different types of development is indicated in Table 3.3.

Table 3.3 *Applicability of ground improvement for different structures and soil types (after West, 1975 and Mitchell, 1981)*

Type of structure	Structure	Permissible settlement	Loading intensity/ bearing pressure (kPa)	Probability of advantageous use of ground improvement methods		
				Loose granular soils	Inert fills	Weak alluvial deposits
Office/ domestic (frame or load bearing construction)	High rise (> 6 storeys)	Low 25 to 50 mm	High 300+	High	Low	Unlikely
	Medium rise (3 to 6 storeys)	Low 25 to 50 mm	Medium 200	High	Good	Low
	Low rise (1 to 3 storeys)	Low 25 to 50 mm	Low 100 to 200	High	High	Good
Industrial	Large span portal, cranes, heavy machines, silos, chemical plant	Low (differential settlement critical)	Variable High local concentrations up to 400	High	Low	Unlikely
	Framed warehouses and factories	Medium	Low 100 to 200	High	High	Good
	Covered storage, storage rack systems, production areas	Low to medium	Low Up to 200	High	High	Good
Others	Road embankments	High	High Up to 200	High, if required at all	High, if required at all	High
	Open storage areas	High	High Up to 250	High, if required at all	High	High
	Storage tanks	Medium to high (differential settlement important)	High Up to 250	High, if required at all	High	High
	Effluent treatment stations	Medium (differential settlement important)	Low Up to 150	High, if required at all	High	High

Any environmental effects of the treatment process, including noise, vibration and pollution, need to be identified. For those treatment methods that are both technically appropriate and environmentally acceptable, cost is likely to have a crucial role in the selection process. The final decision on the adoption of a particular treatment method will require a balanced judgement considering all the advantages and disadvantages of potential treatment methods.

It is unlikely that the treatment of marginal ground and brownfield sites will ever be wholly free from the risk that treatment could be less than fully effective. Risk needs to be managed and the extent, therefore, to which the parties involved in the ground treatment bear the burden of that risk is important. While the risks should not be concealed or glossed over, an unwarranted over-emphasis on risk could form a barrier to the adoption of cost-effective treatment methods. A risk assessment should include consideration of the consequences of inadequate performance of the treated ground, their acceptability or otherwise, and potential remedial measures and costs.

3.4.2 Remedial processes

Treatment methods are introduced in Chapter 6, but prior to examining engineering properties of the ground and their measurement in chapters 4 and 5, it is helpful to have an understanding of the different forms of remedial action that are used in ground treatment, namely:

- densification
- physical reinforcement
- chemical modification.

Particular processes may be in more than one of the above categories. For example, vibrated stone columns can reinforce the ground but may also densify it; similarly lime columns can both reinforce and chemically modify the soil.

Densification

All those types of treatment in which the soil is compressed can be classified under the heading of densification, which includes both compaction and consolidation methods. This is the simplest form of remedial action, which is directly related to deficiencies in the ground associated with high voids ratio and porosity. Methods of *in situ* compaction include vibro-compaction and dynamic compaction. Densification of a non-engineered fill can be achieved by excavation and re-compaction in thin layers.

Consolidation methods include pre-loading with a surcharge of fill together with, where necessary, the installation of vertical drains to speed the rate of consolidation. In some situations, consolidation may be achieved by temporarily lowering the ground water table with consequent increase in effective stress. Densification methods applied from the ground surface, such as pre-loading with a surcharge of fill and dynamic compaction, are suitable for widespread treatment of relatively large areas. They are not normally appropriate or economical to treat a small area under the strip footings of a low-rise building, but they have the advantage of flexibility in the location of buildings after treatment has been completed. Rapid impact compaction can offer effective technical solutions for smaller sites and is able to treat specific foundation areas. Subsurface methods of densification, such as vibro-compaction, are more localised in their effects.

Physical reinforcement

With this approach, the soil is strengthened by the inclusion of reinforcing elements. Such methods can be classified according to the nature and orientation of the reinforcing element and the method of installation. Vibro stone columns and vibro concrete columns are forms of vertical reinforcement which can be used to reinforce the ground locally under small footings and floor slabs without the need to treat large areas as is usually the case with densification methods such as dynamic compaction and pre-loading with a surcharge of fill. Although this localisation of treatment makes the methods cost-effective, there are drawbacks. It can be difficult to design appropriate foundations for extensions to houses that have been founded on vibro stone columns where it is not feasible to install stone columns for the extension. The same problem would occur with piling.

Various types of geosynthetics, such as geogrids, may be used as horizontal reinforcing elements within recompacted ground. They can be combined with vertical reinforcement systems to support widespread loads such as large floor slabs or embankments. For example, the toll plaza for the Second Severn Crossing involved the construction of an embankment over highly compressible soils and the foundation system comprised vibro concrete columns and a load transfer platform of granular fill incorporating low strength geogrids (Maddison *et al*, 1996). This is described in Appendix A6.

Chemical modification

This might be considered to be the most radical approach to ground treatment as the nature of the soil is modified. The stabilising agent, which can be a liquid, slurry or powder, is physically blended and mixed with the soil. Soils have been treated with admixtures of lime, cement or bitumen (Ingles and Metcalf, 1972). Active stabilisers produce a chemical reaction with the soil with consequent desirable changes in the engineering properties. Lime and cement may react chemically with the soil whereas bitumen only acts as a binder. This type of *in situ* treatment has been used in road construction for many years to treat wide areas of ground. Mixing is achieved by suitable earthmoving machines, but only very shallow depths of soil are affected.

From the early 1970s, deep soil mixing has been developed in Japan to improve the properties of cohesive soils to considerable depths. Cement or lime is used and depths of as much as 50 m have been treated. Cement is now the primary agent (Toth, 1993). Lime columns are widely used in Scandinavia. Treatment can be localised under the footings of buildings.

4 Engineering properties

> Many engineering properties exert some influence on the behaviour of treated ground and it is important to identify the more critical properties in a given situation. These can include not only shear strength, stiffness, compressibility, permeability and rate of consolidation, which are routinely determined in many pre-treatment geotechnical investigations, but also properties describing creep behaviour, the potential for collapse on inundation and liquefaction potential. The relative importance of these properties in a particular situation will depend on the nature of the treated ground and the type of the structure that will be built on it. Those properties that govern the field performance of the treated ground are identified and described in this section.

4.1 GROUND PROPERTIES AND FOUNDATION DESIGN

Successful design and construction on treated ground requires knowledge of the modified ground behaviour. The type and extent of knowledge that is required will depend on the form of construction and its sensitivity to settlement. The structural designer will need adequate information about ground behaviour, including bearing capacity and load-settlement characteristics. The following can have a major effect on foundation design:

- soil profile and its variability across the site
- ground water level and any fluctuations in level
- soil properties.

A commonly encountered situation in which ground treatment is required comprises a non-engineered fill overlying a soft compressible natural soil. On such a filled site the depth of fill and variations in the depth of fill may have a critical influence on the durability of the ground treatment and the foundation design. The presence or absence of a ground water table is a primary factor in assessing the potential for collapse compression on inundation in the fill. The age of the fill, the presence of voids or large obstructions within the fill and the presence of contamination are all relevant to foundation design.

Ground behaviour is related to the engineering properties of the soil and it is important to identify the more critical post-treatment properties in a given situation. These include parameters, which are determined in many pre-treatment geotechnical investigations:

- shear strength
- coefficient of volume compressibility and other deformation moduli
- permeability and rate of consolidation
- secondary compression and creep.

The aim of the ground treatment may be to improve one or more of these properties. The direct measurement of engineering properties may therefore be required after, as well as before treatment in order to assess the long-term performance of the treated ground. Methods of measurement are described in Chapter 5, but where ground conditions are very variable it will be difficult to characterise ground conditions and, particularly where heterogeneous fills are present, it will be difficult to assign representative values of soil properties.

The following modes of behaviour, which represent latent deficiencies in the load carrying characteristics of the ground, are closely linked to soil properties and are examined in Section 3.3:

- collapse compression on wetting (Section 3.3.2)
- liquefaction (Section 3.3.3)
- chemical instability (Section 3.3.4).

Some index properties are relevant to foundation design:

- particle size distribution
- moisture content
- plasticity indices
- organic content.

The relative importance of these properties will depend on the nature of the treated ground and the type of the structure to be built on it. The differences are emphasised between treatments that produce an overall densification of the ground and treatments that reinforce the ground with discrete inclusions. Some ground treatment processes introduce variations in ground stiffness; for example, vibro stone columns and the surrounding ground are likely to have very different stiffnesses. In such cases the foundation design should make allowance for this non-uniformity in the treated ground.

4.2 INDEX AND CLASSIFICATION PROPERTIES

The observation and measurement of simple index properties make it possible to classify soils in ways that are relevant to their load-carrying characteristics and likely long-term behaviour.

4.2.1 Moisture content and degree of saturation

The moisture content or water content, w, of the soil is the weight of water expressed as a fraction or percentage of the weight of the solid particles. Moisture contents for saturated soils typically range from 20 to 80 per cent. The behaviour of cohesive soils is very moisture dependent.

The degree of saturation, S_r, is the ratio of the volume of water to the volume of pores and is usually expressed as a percentage; $S_r = 100$ per cent indicates a fully saturated soil. For loose, partially saturated, natural soils and fills, the volume of air voids, V_a, expressed as a percentage of the total volume of the soil is also a useful parameter. The degree of saturation can have a major influence on both collapse potential and liquefaction potential.

Shallow foundations will usually rest on partially saturated ground and the behaviour of soils in this condition is a key factor in foundation performance. Soil above the water table is not subjected to a positive hydrostatic pressure and evaporation and evapo-transpiration may lower the pore water pressures sufficiently to draw out water from the voids in the soil near the ground surface. Thus a zone is created, part of which is saturated and part of which is partially saturated, within which the pore pressures are negative. Suction in fine grained soils can result in an apparent cohesion with enhanced resistance to particle movement. However, small perturbations in ground conditions can destroy the suction and this can lead to ground movements. Collapse compression on wetting is described in Section 3.3.

4.2.2 Particle size distribution

Particle size has a strong effect on soil behaviour and provides a useful method of classification. The distribution of particle sizes in a soil is normally expressed in the form of a grading curve, which shows the percentage by weight finer than a given particle diameter.

There are basic differences in behaviour between coarse granular soils and fine cohesive soils. Coarse soils tend to have high shear strength and permeability whereas fine soils generally have lower strength and permeability. The percentage of silt and clay size particles (ie finer than 0.06 mm), F_c, is important as when this percentage is high, (usually taken as $F_c > 35$ per cent) the soil will cease to behave as a coarse soil. With a well graded soil, it may not be immediately obvious whether the behaviour will be that of a coarse or a fine soil. The BS5930 *Code of Practice for site investigations* gives helpful advice on this transition (British Standards Institution, 1999). If the soil deposit is heterogeneous it will also be difficult to represent it by a grading curve.

Soils are described as uniformly graded where all the particle sizes are within a narrow range, well graded where there is a wide and even distribution of particle sizes and gap graded where there is a wide distribution of particle sizes but with a deficiency in a certain range of particle size. Well-graded sands and gravels are generally less permeable and more stable than those with uniform particle size. Fine uniform sand approaches the characteristics of silt.

The behaviour of fine soils is more complex than that of coarse soils and much more moisture content dependent. The mineralogy has an important influence on the properties of clay soils, but is less important for coarse soils. Clay soils exhibit plastic behaviour within a certain range of moisture contents and considerable strength when air dried. Low permeability makes them difficult to compact when wet. When compacted, they are resistant to erosion and are not susceptible to frost heave, but are subject to expansion and shrinkage with changes in moisture. The properties of clay are influenced not only by the size and shape of the particles, but also by their mineral composition and chemical environment.

The particle size distribution of a soil will affect the suitability and choice of ground treatment technique and the subsequent long-term performance of the treated ground. In general, granular soils are more receptive to methods of surface compaction such as dynamic compaction (Section 7.3). Vibro-compaction (Section 7.2) can be carried out to considerable depths when required and is designed for use within a particular range of sands and gravels with limited fines content. Fine soils may be improved by processes such as pre-loading associated with the installation of vertical drains (Section 8.3) or vacuum pre-loading (Section 8.5).

4.2.3 Plasticity indices

In this context, the term plasticity describes the response of a fine-grained soil to changes in moisture content. The plasticity of a soil is related to the amount of water required to be added to change the consistency of the soil from hard and rigid to soft and pliable. The wetter end of the plasticity range is described by the liquid limit, w_L, and the drier by the plastic limit, w_P. The liquid limit is the moisture content beyond which the soil exhibits "liquid" behaviour; conversely, the plastic limit is the moisture content below which the soil loses its intact behaviour and begins to break up into discrete pieces.

The plasticity of a soil is assessed from its plasticity index, I_p,

$$I_p = w_L - w_P \qquad (4.1)$$

This index may also be used to assess the susceptibility of a clay soil to shrinkage and swelling. As a general rule, the greater the plasticity index the greater the potential for volume change. BRE Digest 240 (Building Research Establishment, 1993c) classifies soils with a plasticity index greater than 40 per cent as having high volume change potential. Treatment of such soils with, for example, vibro stone columns will not alleviate the potential for volume change. Problems for low-rise construction on shrinkable clays can normally be addressed by suitable foundation design.

Some aspects of the behaviour of clay soils may be determined from the value of the natural moisture content relative to the liquid and plastic limits. The liquidity index, I_L, is defined as:

$$I_L = (w - w_P)/(w_L - w_P) \qquad (4.2)$$

At moisture contents between w_P and w_L, the liquidity index gives a simple indication of the position in the plasticity range at which the soil lies. The undrained shear strength at the plastic limit ($I_L = 0$) is of the order of 100 times greater than that at the liquid limit ($I_L = 1$). At moisture contents below the plastic limit I_L is negative.

4.2.4   Compactness

The density or compactness of a soil has a major influence on its behaviour. A dense soil generally will have superior engineering properties to the same soil in a loose condition; thus many forms of ground treatment are essentially methods of densification.

Compactness may be described by terms such as voids ratio, e, porosity, n, dry density, ρ_d, and dry unit weight, γ_d. In assessing field behaviour, it is helpful to relate the *in situ* degree of compactness to that determined in standard laboratory compaction tests, which are described in the next section of the report.

The degree of packing in coarse soils, such as sand and gravel or fill, can be described by the density index, I_D, sometimes known as relative density. Density index provides a useful way of relating the *in situ* dry density of a granular fill, ρ_d, to the limiting conditions of maximum dry density, $\rho_{d\,max}$, and minimum dry density, $\rho_{d\,min}$:

$$I_D = \frac{(\rho_d - \rho_{d\,min})}{(\rho_{d\,max} - \rho_{d\,min})} \frac{(\rho_{d\,max})}{(\rho_d)} \qquad (4.3)$$

Treatment methods such as vibro-compaction and dynamic compaction might typically increase I_D from 0.5 to 0.8. Although density index is an important parameter for coarse soils and strongly influences compressibility, shear strength and vulnerability to liquefaction, its numerical value depends on three densities, which are difficult to determine accurately and the calculated values are liable to significant errors.

The amount of densification that can be achieved for a given compactive effort on a clayey soil is a function of moisture content. Laboratory test results plotted as the variation of dry density with moisture content can be used to identify the two parameters of prime interest: the maximum dry density and the corresponding optimum moisture content. Relative compaction is the ratio of the *in situ* dry density to the maximum dry density achieved with a specified degree of compaction in a standard laboratory compaction test.

4.3 SHEAR STRENGTH

The strength of soil can be described by its shear strength, which is the maximum shear stress that can be sustained by the soil under a given confining stress. Although most geotechnical problems that occur on treated ground concern settlement and volume changes in the ground, shear strength can also be important because it controls the bearing capacity and has a major effect on slope stability, earth pressure and side friction on structures and piles.

There is usually an approximately linear relation between the shear stress at failure, τ_f, and the normal effective stress, σ', which can be expressed as:

$$\tau_f = c' + \sigma' \tan \phi' \tag{4.4}$$

where c' is the cohesion intercept and ϕ' is the angle of shearing resistance.

A coarse soil in a dense state will generally exhibit a greater maximum or peak effective stress shear strength than the same soil in a loose condition, although this effect is suppressed at large normal stresses. When the dense coarse soil is strained beyond the peak strength, there will be a fall in strength to the constant volume or critical state strength, ϕ'_{cv}. This constant volume strength is similar to the maximum strength of the soil in a loose condition where little or no post-maximum fall in strength will occur. For a well graded rockfill where the parent rock has at least moderate strength and which is without excessive fines, ϕ'_{cv} is unlikely to be smaller than 35°. For sands, ϕ'_{cv} typically ranges from 30° to 35°. Ground treatment that densifies a coarse soil should increase its peak shear strength.

Some clay soils exhibit strain softening, commonly referred to as brittle behaviour; as the soil is strained it reaches a maximum strength with a subsequent fall in strength. At extremely high strain levels, for example on the rupture surface of a landslide, the strength of a clay soil may fall to its lowest possible level, the residual strength, which is below ϕ'_{cv}. A clay soil may be considered to have at least three strengths: peak, constant volume and residual, depending on how much deformation has occurred. Treatment of a fine soil by pre-loading is likely to increase the peak shear strength. For low permeability clay soils, the undrained shear strength, c_u, is an important parameter. It is a function of moisture content.

Reinforcement of soft clay using, for example, vibro stone columns (Section 9.2) or soil mixing (Section 9.4) results in a composite system with increased shearing resistance. However, disturbance during column formation can lead to a reduction in strength.

4.4 COMPRESSIBILITY

Buildings and civil engineering structures apply loads to the ground, increasing effective stresses and inducing strains in the soil. An increase in mean effective stress will cause a reduction in volume, the magnitude of which will depend on the compressibility of the soil. An increase in shear stress will cause shear strains, the magnitude of which will depend on the shear modulus of the soil.

The principal type of ground movement in these situations is settlement and the behaviour of soil under laterally confined conditions is of particular importance. The stress-strain behaviour under confined compression is normally expressed in terms of the coefficient of volume compressibility, m_v, which can be expressed in terms of the vertical strain increment, $\Delta\varepsilon_v$, produced by an increment of applied stress, $\Delta\sigma_v$:

$$m_v = \Delta\varepsilon_v / \Delta\sigma'_v \qquad\qquad (4.5)$$

Sometimes the same form of behaviour is described by the constrained modulus, D:

$$D = 1/\,m_v = \Delta\sigma'_v\,/\Delta\varepsilon_v \qquad\qquad (4.6)$$

The deformation under confined compression of both fine and coarse soils can be expressed in terms of m_v or D. However, major differences between the deformation of the granular soil and a typical clay deposit are the magnitude and rate of the volume change; typically the change in voids ratio is an order of magnitude smaller for the granular soil.

A fundamental difference between the deformation behaviour of fine-grained soils (such as clay), and coarse-grained soils (such as sand and gravel) is related to permeability. In a saturated clay soil, the soil grains only come closer together, in response to an increase in applied loading, as water is squeezed out of the soil. However, this change in moisture content occurs slowly due to the low permeability of the soil. Consequently, the initial effect of applying a load to a saturated clay is to increase the pore pressure, without affecting the effective stresses within the soil. The flow of water out of the affected area is controlled by the hydraulic gradient generated by these pore pressures and the drainage conditions at the soil boundaries. As the water flows away from the loaded area, the pore pressures dissipate and the soil compresses; this process is known as primary consolidation.

If a load was applied to a saturated coarse soil totally surrounded by an impermeable barrier, a similar effect would be produced; the applied load would be carried by the increase in pore pressure without affecting the effective stresses. While such conditions can be simulated in the laboratory, the drainage conditions and dissolved gases in the pore water that exist in the ground allow the water to flow and/or compress instantaneously. Consequently, coarse soils do not have a time-dependent consolidation of the type observed in clay soils.

A clay soil will have consolidated as the result of burial during its geological history. Because lateral strains are prevented by the adjacent soil, which is equally loaded, the boundary conditions are similar to those that exist in the oedometer (see Section 5.2). Where soils have been extensively eroded, the near-surface deposits once existed at depths of several hundred metres. These deposits have pre-consolidation pressures that are far greater than their current overburden pressure and are described as heavily over-consolidated.

Soils that have not experienced higher pressures than the current overburden are described as normally consolidated. However, the surface crust of these soils is often over-consolidated as a result of fluctuations in ground water level or desiccation.

Table 4.1 presents a classification of compressibility for clay soils. The soils that would most commonly require ground treatment are in the categories of high compressibility and very high compressibility. Normally consolidated alluvial clays would usually be in the former category and organic alluvial clays and peats may be in the latter category.

Table 4.1 *Classification of compressibility of clays (after Tomlinson, 1995)*

Qualitative description	Coefficient of volume compressibility, m_v (m²/MN)	Constrained modulus, D (MPa)
Very low compressibility	Below 0.05	Above 20
Low compressibility	0.05 – 0.10	10 – 20
Medium compressibility	0.10 – 0.30	3.3 – 10
High compressibility	0.30 – 1.50	0.67 – 3.3
Very high compressibility	Above 1.50	Below 0.67

Ground treatment that takes the form of a temporary pre-loading will effectively over-consolidate the soil. This can increase the constrained modulus of the soil substantially. Table 4.2 shows typical values of constrained modulus for fills appropriate to relatively small foundation loads (Charles, 1993). While an increase in I_D from 0.5 to 0.8 might increase D by a factor of 2, pre-loading has a much greater effect.

Table 4.2 *Typical values of constrained modulus (D) appropriate to foundation loading*

Fill type	D (MPa)
Sandy gravel fill (I_D = 0.8)	50
Sandy gravel fill (I_D = 0.5)	25
Sandy gravel fill (preloaded)	200
Sandstone rockfill (I_D = 0.8)	12
Sandstone rockfill (I_D = 0.5)	6
Sandstone rockfill (preloaded)	40
Colliery spoil (compacted)	6
Colliery spoil (uncompacted)	3
Clay fill (I_P =15%, I_L = 0.1)	5
Old urban fill	4
Old domestic refuse	3
Recent domestic refuse	1

Other parameters such as Young's modulus, E, and Poisson's ratio, v, may be used to describe the stiffness of the soil. For an isotropic, linear elastic material:

$$D = 1/m_v = E (1-v)/([1-2v][1+v]) \tag{4.7}$$

A Young's modulus derived from m_v would not, of course, describe the behaviour of the soil under the stress conditions imposed in the shear stage of a triaxial compression test.

4.5 PERMEABILITY AND RATE OF CONSOLIDATION

Many ground problems are associated with the flow of water through the ground. The ability of water to flow through soil is dependent on the size and distribution of the pore spaces between the solid particles and on how these alter as the soil deforms. Water will flow through the pores of a soil if there is a hydraulic gradient, i, which is defined as the loss of head per unit distance along a flow-line. For one-dimensional flow, it is normally assumed that the velocity of the pore water flow is proportional to the hydraulic gradient. This linear relationship between hydraulic gradient and rate of flow is referred to as Darcy's law and the coefficient of proportionality is known as the coefficient of permeability k of the soil. The rate of flow, Q, through a sample of soil with a cross-sectional area, A, is:

$$Q = kiA \qquad\qquad (4.8)$$

The flow of water exerts a force on the soil. Where the flow of water is upwards, the force can exceed the weight of the soil so that there is no contact between particles and the effective stresses are reduced to zero. For granular soils, the strength is effectively reduced to zero and the soil is described as quick (as in quicksand, for example). The condition is also known as piping as it can result in the weakened material flowing. Quick conditions can occur wherever the upward seepage force exceeds the submerged unit weight of the soil.

Table 4.3 presents a classification of soils according to their permeability. A vast range of permeabilities is found in commonly occurring soils from gravels in the high permeability category to clays in the practically impermeable category.

Table 4.3 *Classification of soils according to their permeability (after Terzaghi and Peck, 1948)*

Degree of permeability	Coefficient of permeability, k (m/s)
High	$> 10^{-3}$
Medium	$10^{-3} - 10^{-5}$
Low	$10^{-5} - 10^{-7}$
Very low	$10^{-7} - 10^{-9}$
Practically impermeable	$< 10^{-9}$

Some typical ranges of permeabilities are given for natural soils in Table 4.4 (Coduto, 1999) and for fills in Table 4.5 (Charles 1993).

Table 4.4 *Typical ranges of permeability for natural soils (after Coduto, 19099)*

Fill type	k (m/s)
Clay	$10^{-8} - 10^{-12}$
Silt	$10^{-5} - 10^{-10}$
Clayey sand	$10^{-4} - 10^{-6}$
Silty sand	$10^{-4} - 10^{-5}$
Fine sand	$10^{-3} - 10^{-5}$
Clean coarse sand	$10^{-2} - 10^{-4}$
Sand-gravel mixtures	$10^{-1} - 10^{-3}$
Clean gravel	$1 - 10^{-2}$

Table 4.5   *Typical ranges of permeability for fills (after Charles, 1993)*

Fill type	k (m/s)
Clay fill (compacted)	$10^{-8} - 10^{-10}$
Sand fill	$10^{-2} - 10^{-4}$
Rockfill (well graded)	$10^{-2} - 10^{-5}$
Colliery spoil (coarse)	$10^{-3} - 10^{-8}$
Pulverised fuel ash	$10^{-6} - 10^{-8}$
Refuse landfill	$10^{-3} - 10^{-5}$

The effect of permeability on the rate of dissipation of excess pore water pressures and associated primary consolidation of clay soils is mentioned in Section 4.4. For many practical problems Terzaghi's theory of one-dimensional consolidation is adequate. The proportion of the total consolidation that has occurred after a given period of time t has elapsed since the application of the load is related to a non-dimensional time factor, T_v :

$$T_v = c_v t / d^2 \qquad\qquad (4.9)$$

Where d is the length of the drainage path and c_v is the coefficient of consolidation:

$$c_v = k / (m_v \gamma_w) \qquad\qquad (4.10)$$

where γ_w is the unit weight of water. It should be recognised that the interpretation of oedometer test results with respect to compressibility is usually more reliable than with respect to time-dependent behaviour. In some situations, predictions of field settlement rates of fine soils, using values of c_v determined on small samples in standard laboratory tests, can greatly overestimate the time required for primary consolidation. Specimen selection favours samples composed of the more clayey material in the deposit because sandy material is more difficult to sample. Also, the length of the drainage path d may have been overestimated where the presence of thin horizontal sandy seams within the soil deposit has not been identified. One approach to this problem is to calculate c_v from laboratory measurements of compressibility and field determinations of permeability.

A number of other important effects and processes are influenced by water movement and permeability, including:

- collapse compression of loose, unsaturated, natural soils and fill (Section 3.3.2)
- liquefaction of some saturated, granular soils and fill (Section 3.3.3)
- loss of ground due to erosion of fine particles from permeable soils
- migration of leachates and contamination.

Uncontrolled seepage can cause the migration of fine particles. Where water flows out of a fill, for example, local instability may occur at the exit point. Where water is flowing through a fill that is susceptible to erosion, the process can be prevented or controlled by protection of the fill with filters, which halt the loss of fine material.

Further information on permeability and seepage can be found in CIRIA Report C515 (Preene *et al*, 2000).

CREEP AND SECONDARY COMPRESSION

In many circumstances long-term movements are of most interest. Creep and secondary compression can cause long-term settlement, which occurs without any change in effective stress or moisture content in the soil.

Many fills show a linear relationship between settlement and the logarithm of the time that has elapsed since the fill was placed. Sowers *et al* (1965) observed this behaviour in embankment dams built of dumped rock fill and described the behaviour in terms of a logarithmic creep compression parameter, α , such that:

$$\alpha = \Delta s / (H \log [t_2 / t_1]) \tag{4.11}$$

where Δs is the settlement of an embankment of height H between times t_2 and t_1 since construction. Values of α ranged from 0.2 to 1.0 per cent; the smaller values were associated with rock fill where there was some compaction. Creep compression can occur due to the weight of the structure built on the fill as well as due to self-weight of the fill.

Figure 4.1 shows the settlement of a 73 m high heavily compacted sandstone and mudstone rock fill embankment. The settlement of the crest of the embankment has been measured over a 20-year period and there is an approximately linear relationship between settlement and the logarithm of time that has elapsed since the end of construction corresponding to $\alpha = 0.17$ per cent.

The response of a saturated clay soil to loading differs considerably from that of a granular soil. The load initially induces excess pore pressures in the clay soil and there may be little immediate settlement. Time-dependent movements of two types then occur; primary consolidation and secondary compression. Primary consolidation slowly continues until all the excess pore pressures have dissipated. This is followed by secondary compression, which continues under conditions of constant effective stress. This secondary compression of saturated clay fills exhibits a linear relationship between settlement and the logarithm of the time that has elapsed since the load was applied. The continuing rate of compression of clay fills in one-dimensional compression can be described by the coefficient of secondary compression, C_α:

$$C_\alpha = \Delta h / (h \log [t_2 / t_1]) \tag{4.12}$$

where compression $\Delta h / h$ occurs between times t_2 and t_1 after the load was applied. Typically C_α for clay fill might be in the range 0.001 to 0.005 (0.1 to 0.5 per cent).

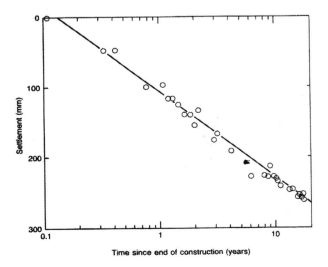

Figure 4.1 *Long-term settlement of 73 m high rock fill embankment (after Charles, 1993)*

Creep and secondary compression movements are likely to be relatively small except for peat and other organic soils. They will not normally be of such magnitude as to necessitate ground treatment, and the number of treatment methods that will substantially reduce this type of movement is limited. With loose fills, treatment methods that pre-load the ground should reduce the creep rate, whereas treatment methods such as dynamic compaction may actually increase the rate of creep for a time.

Where fills contain domestic refuse there will be long-term settlement associated with volume reduction caused by biodegradation of the organic matter as well as physical creep compression. The typical heterogeneity of domestic refuse, see Figure 4.2, will influence the amount and variability of the long-term settlement across a site. Watts and Charles (1999) have described the use of the parameter, α_c, to represent the long-term settlement of landfills due to physical effects and a similar parameter, α_b, for the settlement due to bio-consolidation. The latter process will be the dominant cause of long-term settlement in fill with a high organic content and will be affected not only by changes in effective stress, but also by changes in the biochemical environment. Typically values of α_b may be an order of magnitude greater than α_c.

Table 4.6 summarises constrained moduli prior to treatment and settlement performance subsequent to treatment of some old refuse fills (circa 1960), which were treated either by dynamic compaction or pre-loading with a surcharge of fill. At a site in Redditch, dynamic compaction substantially reduced the compressibility of a refuse fill although treatment did not eliminate long-term settlement. Surcharge treatment on old refuse fills at Redditch and Liverpool induced substantial settlements as the surcharge was placed and it was unnecessary to leave the surcharge in place for a long period. However, surcharging did not eliminate long-term settlement. The use of vibro stone column treatment in biodegradable fill is generally inappropriate but vibro concrete columns and stabilised soil columns may provide useful solutions.

Figure 4.2   *Example of the variability of domestic refuse (courtesy of the Building Research Establishment Ltd)*

Table 4.6   *Properties of refuse fills (after Watts and Charles, 1999)*

Location	Ground treatment	Depth of refuse	Constrained modulus, D (MPa)	Long-term settlement	
				α_c (%)	α_b (%)
Redditch	DC	6.0	6	0.2	2
Redditch	Pre-loading	5.8	2	–	2
Liverpool	Pre-loading	8.0	3	0.2–0.3	6

5   Measurement of engineering properties

> The measurement of engineering properties has a critical role in the successful application of ground treatment. It is important to measure those properties that will determine the success or failure of the treatment. Measurements may be made during successive stages of the project to fulfil different objectives. Various forms of penetration testing are useful in the *in situ* assessment of treated ground and the cone penetration test is often used. Geophysical techniques are also helpful in some situations. Load tests can provide full-scale evidence of ground performance.

5.1   OBJECTIVES

On a site where some form of ground treatment is carried out, the effectiveness of the treatment should be assessed. The crucial issue is usually the long-term performance of the treated ground and particularly post-construction ground movement. Unfortunately for the client, unsatisfactory performance may only become apparent in the long term. In the interim some appropriate measures of improvement are required. The measurement and evaluation of engineering properties are therefore of great importance. Small projects present particular difficulties as the funds available to carry out any form of testing will be very limited; ground treatment may have been chosen for reasons of cost and a requirement for extensive testing could remove the cost advantage.

Measurements of ground properties should be made before treatment, during treatment, immediately following treatment and in the long-term. However, some treatment methods induce an increase in pore pressures within the material being treated and so performance testing conducted during and immediately post-treatment could be misleading. The measurements made at these different stages in the project have different objectives and may markedly differ from each other. Following a review of these objectives, the measurement of the properties of treated ground are examined under the following headings:

- laboratory tests
- *in situ* penetration tests
- geophysical tests
- load tests.

5.1.1   Prior to ground treatment

Measurements of soil properties will usually be made as part of the initial site investigation. This investigation will have a number of objectives including the development of a three-dimensional stratigraphic model of the ground conditions, a ground water model, the measurement of soil properties of each stratum and the prediction of ground behaviour for the type of structure to be built on the site. The ground investigation should also provide the engineering parameters needed to produce a practical and economic foundation design or ground improvement strategy to allow the site to be successfully developed.

The site investigation process is described in Section 3.2. The site investigation, of which the measurement of soil properties forms part, should lead to the diagnosis of any deficiencies in the ground, which require to be remedied, and the identification of suitable treatment methods. *In situ* testing may be carried out at this stage so that a comparison with post-treatment testing can provide a measure of the improvement that has resulted from ground treatment.

5.1.2 During ground treatment

With some forms of ground treatment, valuable data on soil behaviour can be obtained as the treatment progresses. This makes it possible to use an observational approach in which the ground treatment process is modified as required by the behaviour monitored during treatment. The extent to which this can be implemented will depend on the type of ground treatment, the sophistication of the monitoring and contractual arrangements. It should be recognised that some forms of ground treatment induce high pore water pressures in the soil being treated and testing carried out during and immediately after treatment could be misleading. The principles and applications of the observational method in ground engineering are described in CIRIA Report 185 (Nicholson *et al*, 1999). It is essential that contingency plans are defined in advance to deal with observational findings that deviate significantly from design assumptions and that there is a clear line of decision-making to implement them.

The recording of data during the installation of vibrated stone columns provides a great deal of information on both the soil in which the stone columns are being installed and the columns themselves. It can be linked to quality assurance and quality control. The use of a rig-mounted computerised instrumentation system to monitor and record the treatment process can greatly enhance the gathering of reliable data. In rapid impact compaction of a non-engineered fill, the depth to which the compacting foot is driven after each drop of the weight gives a measure of the compactness of the surface layers of the fill. In both these examples the information obtained during ground treatment forms in effect an extension to the pre-treatment site investigation; for example, particularly soft or hard areas of ground can be identified. If this type of information is properly evaluated at the time of treatment, the process can be modified where appropriate as treatment progresses.

5.1.3 Immediately after completion of ground treatment

Testing should be carried out immediately following the completion of ground treatment to assess its effectiveness. This could include any of the techniques used in the initial site investigation such as excavation of trial pits, drilling of boreholes, sampling and laboratory testing. *In situ* testing, such as the standard penetration test, cone penetration test or dynamic probing, may be carried out to give a direct comparison with tests carried out before treatment. An end-product specification may require certain test values to be obtained.

Short-term load tests may be carried out in connection with a performance specification. If carried out on a sufficiently large scale, they can give a direct measure of the response of the treated ground to the magnitude and distribution of loading, which subsequently will be applied by the building or other type of structure.

5.1.4 In the long term

Commercial imperatives to build on the land as soon as treatment is completed, leaves little scope for monitoring prior to construction; and while monitoring of the settlement of buildings can provide invaluable information, it is rarely undertaken. Nevertheless, all the parties concerned in a ground treatment project should appreciate or be made aware of the importance of long-term monitoring and the extra degree of security that it provides.

Every effort should be made to promote long-term monitoring. It provides a record of when movements began and the rate of movement, which helps in the diagnosis of problems and the design of remedial measures. Some resistance to such monitoring comes from the costs involved and a belief that to carry out such monitoring indicates doubt as to the success of the ground treatment. When long-term performance is the critical issue, the installation of instrumentation to monitor settlement and, where appropriate, pore water pressure is recommended.

5.2 LABORATORY TESTS

Soil samples are usually obtained from trial pits and boreholes and may be used for several purposes:

- visual inspection for soil description
- measurement of index and classification characteristics (Section 4.2)
- measurement of engineering properties such as shear strength, compressibility and permeability (sections 4.3, 4.4 and 4.5)
- evaluation of potential for creep, collapse compression and liquefaction (sections 4.6, 3.3.2, 3.3.3).

Tests should be carried out in accordance with BS 1377 *Methods of test for soils for civil engineering purposes* (British Standards Institution, 1990). Additional guidance on laboratory testing is given in the three volume *Manual of soil laboratory testing* (Head, 1998). Requirements for the execution, interpretation and use of geotechnical laboratory tests are provided in *Eurocode 7: Geotechnical design – Part 2: Design assisted by laboratory testing*, DD ENV 1997-2:2000 (British Standards Institution, 2000a). Guidance on the selection of laboratory tests and practical considerations involving the procurement, administration and scheduling of tests has been provided by the Association of Geotechnical and Geoenvironmental Specialists (1998c).

Bulk disturbed samples are adequate for index properties, while "undisturbed" tube samples are required for strength and deformation tests. Even in the best quality tube samples, some disturbance is inevitable. Tube samples are often taken at close intervals in a borehole to ensure that maximum information is obtained from the borehole. Water samples may be taken from both boreholes and trial pits. Block samples can be taken from trial pits. It is difficult to obtain undisturbed samples of most non-engineered fills. Additionally, the variability of soils and materials encountered on sites that may require ground treatment means it is often difficult to obtain representative samples.

Window sampling can provide a fast and cost-effective method of ground investigation (Anon, 1997). The light-weight dynamic window sampler uses special sampling tubes, which are driven into the ground using a high frequency percussion hammer. Most of the dynamic sampling undertaken in the United Kingdom is carried out using tubes with a window slot along one side of the tube. The use of the technique is limited by soil type; it is more suitable for clayey materials than for coarse natural soils and fills.

At depths below 6 m to 7 m progress rates tend to rapidly decrease with increasing depth (Eccles and Redford, 1997). The use of this sampling technique in potentially contaminated ground has been described by Eccles and Redford (1997).

5.2.1 Index and classification tests

BS 1377 Part 2 (British Standards Institution, 1990) specifies standard procedures for carrying out most commonly used classification tests for soils. Index tests include tests to determine moisture content, density, liquid and plastic limits, particle size distribution, particle density (sometimes termed specific gravity) and organic content. Index tests provide an overall engineering classification of the soil and assist in identifying variations within the ground.

Where the natural moisture content of a soil is to be compared with the plastic and liquid limits, the natural moisture content of the particle size fraction used in the plastic and liquid limit tests should be used.

There are two types of test procedure for determining the particle size distribution. Coarser materials, such as sand, gravel, cobbles and boulders, are separated by sieving, and the measured particle size is therefore the size of sieve through which the particle will pass. The finer materials are separated by sedimentation. The particles are dispersed in a fluid and allowed to settle under the influence of gravity and the measured particle size is the diameter of the equivalent sphere of similar density material that would fall at the same rate.

Standard laboratory compaction tests are described in BS 1377 Part 4 (British Standards Institution, 1990). Cohesive soils are normally compacted by multiple blows with a fixed weight rammer, while an electrical vibrating hammer can be used for granular soils.

5.2.2 Strength and deformation tests

Procedures for standard laboratory tests, which can be used to determine deformation and strength parameters, are specified in BS 1377 (British Standards Institution, 1990). It is unwise to rely on single results as materials and laboratory performance vary. It is important that sufficient sets of tests are carried out on appropriately sized specimens of representative material to characterise reliably the shear strength and compressibility parameters of all the soil types affecting foundation performance.

In an oedometer test the soil sample is deformed one-dimensionally, by loading it vertically and at the same time preventing any lateral movement from taking place.

The stress-strain behaviour of a clay soil under confined compression is normally expressed in terms of the coefficient of volume compressibility, m_v, or its inverse, the constrained modulus.

Part 5 of BS 1377 (British Standards Institution, 1990) describes test procedures for the measurement of consolidation properties in the commonly used mechanically loaded oedometers. Porous discs placed in contact with the top and bottom faces allow the sample to drain freely from both faces. More complex testing procedures in hydraulically loaded one-dimensional consolidation cells and triaxial consolidation cells are described in Part 6 of the standard. The reasons why laboratory tests are usually poor at predicting the rate of consolidation in the field have been summarised by Clayton *et al* (1995).

Shear strength may be measured in a shear box or, more commonly, in triaxial compression tests. The simplest form of triaxial test on cohesive soils is the unconsolidated undrained test in which the undrained shear strength, c_u, is measured. This type of testing is described in BS1377 Part 7 (British Standards Institution, 1990).

The effective stress shear strength parameters, c' and ϕ', can be determined from a set of triaxial tests, using a minimum of three tests at different consolidation pressures. The tests may be either consolidated drained with measurement of volume change or consolidated undrained with measurement of pore water pressure. Both types of test are specified in Part 8 of BS1377 (British Standards Institution, 1990). In the consolidated drained test, the axial deformation should be sufficiently slow that pore pressure changes due to shearing are negligible, but sufficiently fast to prevent creep. In the consolidated undrained test, the axial deformation should be at a rate slow enough for adequate equalisation of pore pressures. The period required for a consolidated drained test on a clay soil may be many days especially for larger specimens.

5.2.3   Special properties

It is important that the schedule of tests is relevant to the matters under investigation and so factors such as liquefaction potential, susceptibility to collapse compression on inundation and creep behaviour may need to be investigated. Unfortunately, it is not always possible to obtain reliable data on these aspects of ground behaviour from standard laboratory tests.

Earthquakes induce cyclic loading in soil deposits and cyclic triaxial compression tests are used to investigate resistance to liquefaction. Some guidance on the execution and interpretation of such tests has been given by the Japanese Geotechnical Society (1998). An example of some testing carried out in an investigation of the feasibility of building development on hydraulically placed PFA at Peterborough is given in Box 5.1.

Collapse potential is of particular concern for non-engineered fills and is examined in Section 3.3.2. The measurement of collapse potential has been reviewed by Charles and Watts (1996). While collapse compression on inundation can be determined in laboratory oedometer tests, it is difficult to obtain undisturbed samples of most non-engineered fills.

Box 5.1   *Investigation of liquefaction resistance of PFA (after Humpheson et al, 1991)*

As part of an investigation of the sustainability of hydraulically placed PFA as a foundation for buildings, cyclic undrained triaxial tests were carried out on both undisturbed and resedimented samples of PFA. The results are shown in Figure 5.1 as graphs of cyclic shear stress ratio plotted against the logarithm of the number of cycles causing 5 per cent single amplitude strain. Undisturbed samples were more resistant than silty sand samples.

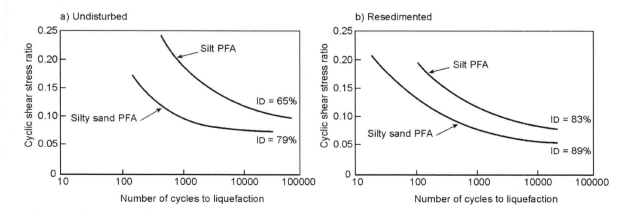

Figure 5.1   *Liquefaction resistance of PFA (after Humpheson et al, 1991)*

5.3   *IN SITU* TESTS

Geotechnical *in situ* testing, in which assessments of soil properties are made either during boring and drilling or from instruments inserted into the ground, has certain advantages over conventional sampling and laboratory testing. Tests can be carried out in soils that are impossible or difficult to sample without using expensive techniques. A larger volume of soil may be tested than is usually practicable in laboratory testing, and this should be more representative of the soil mass. The tests provide a large amount of information quickly and can therefore be relatively cheap. Improvement in strength and stiffness parameters may be assessed using penetration methods (Mitchell, 1986; Mitchell and Brandon, 1998) but difficulties may be encountered with compacted heterogeneous material on brownfield sites (Butcher and McElmeel, 1993).

In situ testing techniques have limitations. Inserting an instrument into the ground will cause a certain amount of disturbance. It is not obvious how much soil is effectively tested and there is limited scope for controlling the stress conditions under which tests are conducted. Push-in devices are not suitable in rock fill or earth fill that contains large particles. Empirical correlations usually have to be applied to obtain engineering properties from test results. Even with the field vane test, which gives a direct measurement of undrained shear strength, the value measured in the test is not normally the value that would be measured in an undrained triaxial compression test.

It is difficult to obtain undisturbed samples of sands and gravels, and consequently, at sites where there are granular soils, geotechnical design parameters are often inferred from *in situ* test results. The evaluation of design parameters for granular soils from *in situ* tests has been reviewed by Jamiolkowski *et al* (1998). They concluded that the assessment of density index and the angle of shearing resistance from cone penetration tests, flat dilatometer tests and standard penetration tests was well established. Table 1.1 of Lunne *et al* (1997) provides guidance on the applicability and usefulness of common *in situ* tests for the determination of particular soil parameters in various soil types.

Tests should be performed in accordance with national or international standards whenever possible. BS 1377: Part 9 (British Standards Institution, 1990) specifies test methods for the standard penetration test (SPT), cone penetration test (CPT), dynamic probing (DP), field vane and plate loading test. BS 5930 (British Standards Institution, 1999) also contains important information on field tests and, in particular, covers permeability testing.

For a number of commonly used field tests, Eurocode 7: *Geotechnical design – Part 3: Design assisted by field testing*, DD ENV 1997-3:2000 (British Standards Institution, 2000b) provides requirements for equipment and test procedures, and reporting, presentation and interpretation of test results.

For most types of ground, field permeability tests yield more reliable data than those carried out in standard laboratory tests. The significance of this for predicting field settlement rates of fine soils has been described in section 4.5. In a field test, a somewhat larger volume of material is tested and this volume of material is more likely to be representative of the soil deposit as a whole, including any thin layers of more sandy material, which might not be present in small laboratory samples. Ground disturbance during sampling is avoided by field testing. Permeability tests can be carried out using standpipe piezometers installed in boreholes as described in BS 5930 (British Standards Institution, 1999). The standpipe piezometers can also be used to monitor ground water levels and piezometric pressures.

5.3.1   Cone penetration test (CPT)

The CPT has gained in popularity in the United Kingdom over the last 20 years because it is quick and its results are reproducible, requiring minimum site preparation. The methods and interpretation of cone penetration testing have been described by Meigh (1987) and Lunne *et al* (1997). The penetration resistance of a cone is measured as it is advanced at a uniform rate into the soil. The basic measurements are the cone resistance, q_c, and the sleeve friction, f_s, from which is derived the friction ratio $R_f = f_s/q_c \times 100\%$. Testing using a CPT truck is shown in Figure 5.2.

The piezocone (CPTU) incorporates a pore pressure sensor and may become the standard version of the CPT. The measurement of pore water pressure can significantly improve interpretation of the test results. Rate of consolidation parameters can be assessed by measuring the dissipation of pore pressure with time after a stop in penetration. Pore pressure response provides an indication of the *in situ* permeability of the ground and enables the presence of thin permeable layers to be identified.

Figure 5.2 *CPT truck (courtesy of Building Research Establishment Ltd)*

Figure 5.3 shows some measurements made by CPTU in hydraulically placed PFA. Layers up to 2 m thick had very low values of q_c (< 0.5 MPa) and R_f (< 1%) and high values of pore water pressure.

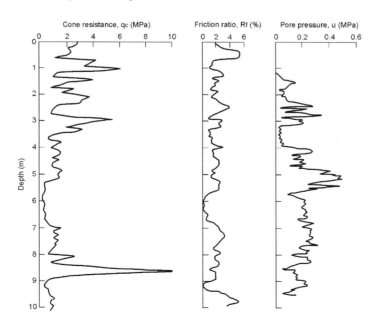

Figure 5.3   *CPTU in hydraulically placed PFA (after Humpheson et al, 1991)*

The CPT can provide information on the following:

- ground conditions between boreholes including the soil profile and identification of soil types

- stratigraphy of soil layers and their homogeneity across the site

- geotechnical properties of the soil; for example some correlations between constrained modulus and q_c can be found in Lunne *et al* (1997) and in DD ENV 1997-3:2000 (British Standards Institution, 2000)

- liquefaction potential of granular soils (Stark and Olson, 1995; Lunne *et al*, 1997).

The cone resistance measured in the site investigation for a housing development at Wythenshawe is shown in Figure 5.4 (Watts *et al*, 1989). Values of q_c of 1 MPa were obtained in a thin peat layer and 20 MPa in a sand/gravel layer. Vibro stone columns were subsequently installed at the site using the wet method (Section 9.2).

The CPT has many advantages. The test is carried out quickly and a borehole is not needed unless there are obstructions or dense granular deposits at the surface to penetrate. A near continuous vertical profile of the soil is obtained as the cone advances and there are correlations for deriving various geotechnical parameters from these measurements. An indication is given of the need for treatment and the suitability of different forms of treatment. For example, certain soils with a cone resistance less than 3 MPa can be highly compressible and vibratory compaction is generally more effective in soil deposits with a friction ratio smaller than 1 per cent (Lunne *et al*, 1997).

The test method has limitations. There are many ground conditions, such as those containing gravel, in which the cone would be liable to be damaged. With even the largest capacity rig, penetration into dense gravels is small and cobbles and boulders can prevent further progress. Systems exist that can overcome some of these problems but add time and cost.

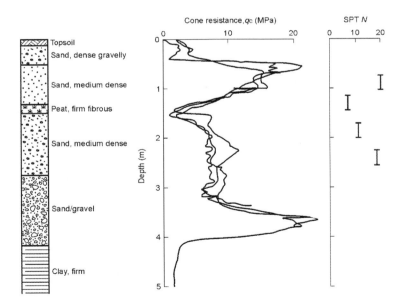

Figure 5.4   *CPT in sand with peat layer (after Watts* et al, *1989)*

There have been many attempts to correlate q_c with SPT N values and much of this data has been summarised by Lunne *et al* (1997). Some site investigation data from a sand site, where both CPT and SPT measurements were made, are shown in Figure 5.4. The ratio q_c/N increases as the mean particle size, D_{50}, of the soil increases but there is a large scatter in the results (Mitchell and Brandon, 1998) and this is indicated in Figure 5.5. In view of this scatter, it is preferable to use a correlation between CPT results and the required engineering property rather than convert the CPT data to equivalent SPT values and then use a correlation between SPT and the engineering property.

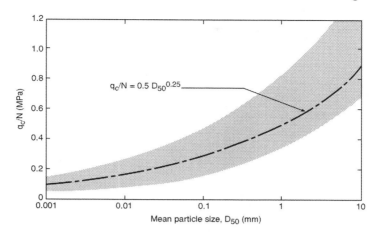

Figure 5.5 *Correlation of CPT-SPT ratio with particle size*

5.3.2 Standard penetration test (SPT)

The SPT gives a measure of the resistance of soil at the base of a borehole to the penetration of a thick walled tubular steel sampler. The methods and use of the test have been described by Clayton (1993). The interpretation of this crude and simple test is empirical but it has probably been the most widely used *in situ* test and there is a large quantity of data empirically relating geotechnical properties to the blow count, N. The test also provides an indicator of the liquefaction resistance of granular soils. The SPT can give useful additional information for little extra cost where boreholes are being drilled to obtain samples.

Many SPT based design correlations were developed using hammers with an efficiency of about 60 per cent. To allow for variations in efficiency, Skempton (1986) has described a method for converting the blow count, N, recorded in the field into a normalised N_{60}. A further normalisation of N to an overburden pressure of 100 kPa, is represented as $(N_1)_{60}$.

Table 5.1 presents a correlation between the density index, I_D, and the normalised blow count $(N_1)_{60}$ for normally consolidated natural sand deposits. This is reproduced from an informative annex to DD ENV 1997-3: 2000 (British Standards Institution, 2000) and is derived from Skempton (1986). For $I_D > 0.35$, $I_D = [(N_1)_{60}/60]^{0.5}$ approximately. For fine sands, the N values should be reduced in the ratio 55/60 and for coarse sands the N values should be increased in the ratio 65/60.

Table 5.1 *Correlation between density index and SPT normalised blow count*

	I_D	$(N_1)_{60}$
Very loose	0 – 0.15	0 – 3
Loose	0.15 – 0.35	3 – 8
Medium	0.35 – 0.65	8 – 25
Dense	0.65 – 0.85	25 – 42
Very dense	0.85 – 1.0	42 – 58

Dynamic probing (DP)

There is some similarity between DP and the SPT but DP is carried out from the ground surface rather than from the base of a borehole, and a near continuous record of the effort required to drive a solid cone through the soil is obtained (Butcher *et al*, 1995). There is no provision for measuring the force on the cone itself, unlike the CPT cone. The test can supplement conventional sampling and other more complex penetration tests, and is particularly useful in delineating areas of weak soils overlying stronger strata and in quickly assessing the variability of the soil conditions. It is less successful in identifying differences in soil type; for example, the blow count in loose granular soil is similar to that in stiff clay. It provides a simple inexpensive method using lightweight equipment for obtaining a rapid indication of the density index and depth profile of relatively soft or loose deposits. Figure 5.6 shows dynamic probing in miscellaneous fill.

Dynamic probing should be used in conjunction with boreholes if estimated soil parameters are to be interpreted from the results. Figure 5.7 illustrates the way in which dynamic probing was used to assess the improvement in properties of an ash fill treated by rapid impact compaction.

Figure 5.6   *Dynamic probing in miscellaneous fill (courtesy of Building Research Establishment Ltd)*

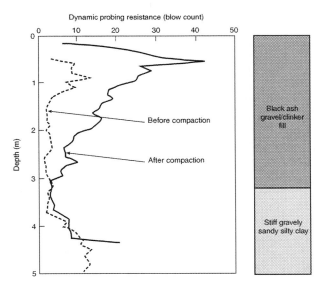

Figure 5.7   *Improvement of old ash fill (after Watts and Charles, 1993)*

5.3.4 Pressuremeter (PMT)

This cylindrical device has a flexible membrane, which imposes a uniform pressure on a borehole wall. The applied pressure and the resulting deformation of the membrane are measured. The outward radial horizontal deformation is no more than a few millimetres. Parameters associated with soil stiffness and strength, and also in certain cases *in situ* stress, may be calculated from the results of the test.

The pre-bored pressuremeter, which is commonly of the Menard type, is inserted into a borehole and the test procedure is relatively simple. The self-boring pressuremeter, which incorporates a drilling head with a drill cutter, is a more sophisticated instrument requiring considerable skill in its operation. The self-boring pressuremeter should cause much less disturbance than the pre-bored pressuremeter, however it is a more complex and therefore more expensive method of testing.

Methods of testing and the interpretation of results have been described by Mair and Wood (1987) and Clarke (1995). The PMT is included in DD ENV 1997-3:2000 (British Standards Institution, 2000).

5.3.5 Flat dilatometer test (DMT)

This device (Marchetti, 1980) consists of a 15 mm thick flat stainless steel blade measuring approximately 0.22 m × 0.1 m. One face of the blade has a recess to take a 60 mm diameter flexible steel diaphragm, which is flush with the surface of the blade. The membrane is inflated by gas pressure. The blade is pushed into the ground in 0.2 m increments by means of rods and at each test position two pressure measurements are made:

- p_0, pressure required to begin to move the diaphragm
- p_1, pressure required to move the diaphragm centre 1 mm into the soil within a time interval of between 15 and 30 seconds.

Conventional geotechnical parameters are not measured directly, but there are empirical correlations with *in situ* horizontal stress, undrained shear strength of clays and deformation moduli. Powell and Uglow (1988) assessed the usefulness of the test for soils in the United Kingdom. In many cases the CPT is likely to be preferred due to the greater continuity in the measurement profile and the greater speed of testing.

5.3.6 Field vane test

The test was developed to measure the undrained shear strength of very soft and sensitive marine clays and provides a direct method of determining the *in situ* undrained shear strength, c_u, of soft to firm clays. The vane is cruciform in shape, comprising four blades set at right angles to each other. BS 1377: Part 9 (British Standards Institution, 1990) specifies vanes 100 mm or 150 mm long depending on soil strength and with length-to-width ratios of 2.

With the borehole apparatus, the vane is lowered into the borehole on extension rods. The vane is pushed a short distance, at least three times the borehole diameter or twice the vane length, whichever is the greater, into the undisturbed soil below the bottom of the borehole.

With the penetration vane test equipment, the vane and a protecting shoe are jacked or driven into the ground. At the required depth the vane is advanced at least 0.5 m from the protective shoe into the undisturbed soil. A torque measuring device is connected to the top of the upper extension rod. The vane is slowly rotated and the maximum torque required to shear the soil is used to calculate the undrained shear strength. Continued rotation of the vane allows calculation of the remoulded undrained shear strength.

Although the test gives a direct measurement of undrained shear strength, this is not necessarily the same as would be measured in an undrained triaxial test, nor is it the same as the true *in situ* strength. Bjerrum (1973) discussed the factors that can be applied to correct vane shear strengths to obtain the field strength for design purposes. DD ENV 1997-3:2000 (British Standards Institution, 2000) gives some examples of correction factors based on plasticity index and vertical effective stress.

5.4 GEOPHYSICAL TESTS

There has been a steady growth in the application of geophysical techniques to civil engineering studies. CIRIA report C562 *Geophysics in engineering investigations* will provide more detailed information on the subject. Geophysical testing has some major advantages:

- fieldwork is relatively rapid and, with modern data logging facilities and processing software, the results can be presented very quickly
- non-intrusive surveys can be carried out from the ground surface
- representative values of soil parameters can be measured.

In general, the techniques should be used in conjunction with conventional procedures and not as an alternative. The results require careful correlation with borehole data. Although methods such as ground probing radar and electrical resistivity are sometimes used, in many situations seismic methods are the most useful geophysical techniques.

5.4.1 Seismic methods

A seismic wave transmits energy by vibration of soil particles, but the wave velocity is quite distinct from the velocity at which individual particles oscillate; in compression waves the soil particles vibrate in the direction of wave propagation and in shear waves the particles vibrate in a direction perpendicular to the direction of wave propagation. Rayleigh waves are distortional stress waves that propagate near to the ground surface.

The dynamic shear modulus, G_{max}, and the dynamic constrained modulus, D_{max}, can be calculated from, respectively, the shear wave velocity, V_S, and the compression wave velocity, V_P:

$$G_{max} = \rho V_S^2 \tag{5.1}$$

$$D_{max} = \rho V_P^2 \tag{5.2}$$

where ρ is the soil density.

Typically $V_S = 1.05 V_R$ where V_R is the velocity of the Rayleigh waves. The velocity of compression waves in water is about 1500 m/s, which is considerably higher than in most soil and fills. The shear wave velocity in water is zero. For a saturated clay fill, the dynamic constrained modulus will be related to undrained behaviour.

Seismic methods may be used for a wide variety of purposes:

- measuring soil properties and their spatial variation
- determining the location and depth of boundaries such as bedrock beneath overburden and the depth of the ground water level

- identifying localised features including defects and voids in the treated ground.

When the methods are used to measure soil properties, the small-strain geophysical properties need to be related to the properties at what may be larger strains in the ground under foundations and adjacent to engineering structures (Matthews et al, 2000).

Seismic waves can be generated by some form of impact or vibration source on the ground surface or in a borehole. Shear waves can be produced by striking a steel plate fixed to the ground with a blow parallel to the ground surface. Measurements of wave velocity can be made from the ground surface or from boreholes. Seismic detectors are known as geophones. Using cross-hole tomography, the distribution of the wave velocities over a plane section within the ground can be determined.

In refraction surveys the time is measured for waves to pass from the source to a number of geophones placed on the surface of the ground at different distances from the source. The analysis of refraction measurements requires the assumption of an increasing velocity with depth and a relatively consistent soil profile along the survey line. The seismic refraction technique is commonly used to delineate the depth to bedrock.

There are two forms of surface wave source in use:

- impact sources such as a hammer or drop weight produce a transient pulse
- vibrators produce continuous waves.

Where an impact source is used, the data is usually analysed using the spectral analysis of surface waves method (SASW). Vibrator sources have been widely used with the continuous surface wave system (CSW).

5.4.2   SPECTRAL ANALYSIS OF SURFACE WAVES (SASW)

The spectral analysis of surface waves is a non-intrusive seismic test for determining wave velocity profiles (Matthews et al, 1996). The method uses a hammer or other type of impact as an energy source. The test is based on the principle that the depth of the soil profile sampled by surface waves varies with frequency and hence wavelength. In most soils the Rayleigh waves travel at a depth of between one half and one third wave length below ground surface. Thus Rayleigh waves of different wavelengths propagate at different depths and if the stiffness of the soil varies with depth, surface waves of different wavelengths will propagate at different velocities. The method uses the spectral analysis of the propagating Rayleigh wave to determine the frequency wavelength dispersion.

The SASW method has become an important method for evaluating the variation with depth of stiffness moduli at small strains. It has also been used to:

- detect underground obstacles and cavities (Ganji et al, 1997)
- evaluate liquefaction potential (Andrus et al, 1998)
- evaluate the effectiveness of dynamic compaction (Kim and Kim, 1997)
- evaluate the effectiveness of vacuum consolidation (Haegeman and Van Impe, 1998)
- evaluate the effectiveness of vibro treatment methods.

5.4.3 Continuous surface wave method

The continuous surface wave system is also used on the ground surface and makes use of Rayleigh waves. The wave source is a frequency controlled vibrator, which makes it possible to derive the relationship between Rayleigh wave frequency and wavelength and hence calculate velocity as a function of depth (Abbiss, 1981). The method has been used at an old chalk quarry at Swanscombe, which had been infilled with Thanet sand, to monitor the improvement in stiffness resulting from the installation of vibro stone columns. Butcher and McElmeel (1993) described the use of CSW to determine the depth and degree of effectiveness of rapid impact compaction in the treatment of a loose building waste fill. Matthews *et al* (1996) have discussed the relative merits of the two surface wave methods. The CSW method requires a relatively costly energy source but has the advantage of good frequency resolution.

5.4.4 WAK test

Briaud *et al* (1989) and Briaud and Lepert (1990) have proposed a test, the WAK test, in which a dynamic compaction weight is hit with an instrumented sledge hammer while it rests on the ground after being dropped by a crane. Geophones on the weight record the response and give a measure of the stiffness of the fill.

5.5 LOAD TESTS

Direct measurements of settlement characteristics through field loading tests form an important part of a testing programme. The scale and duration of such tests vary enormously, depending on the objectives of the test and the type of ground being tested.

Small plate loading tests are simple and inexpensive, but the loaded plate will only apply significant stresses to a very shallow depth. Plate loading tests are commonly carried out on vibro stone columns as a form of quality control and to determine the bearing capacity of an individual stone column. The load is applied to the plate by jacking against the reaction provided by a heavy vehicle; the immediate settlement can be measured as the load is increased. The test merely loads one stone column and gives no indication of the composite behaviour of the ground reinforced by the stone columns. The test provides little, if any, information about the long-term settlement characteristics of the treated ground under working load, because it is completed in a matter of hours or less.

The pressure to apply and the area over which it should be applied will depend on the foundation load and widths. The length of time the test should be maintained is important as the results will have to be extrapolated to predict long-term foundation settlement. The number of tests required at a particular site will depend on the size of the site, the nature of the development and the variability of the ground.

Figure 5.8 shows a pad loading test, which was used to evaluate the effectiveness of dynamic compaction. The load was applied to the pad by jacking against the reaction provided by kentledge.

Figure 5.8 *Pad loading test (courtesy of Building Research Establishment Ltd)*

Generally, the principal objective of testing is to estimate the long-term settlement of the treated ground under working load. The type of test that is required is one in which the load can be kept constant over a comparatively long period and this can be done most simply by the direct application of dead weight. An appropriate specification for this is given in BS 1377 Part 9: *Determination of the settlement characteristics of soil for lightly loaded foundations by the shallow pad maintained load test* (British Standards Institution, 1990).

The party wall of a typical modern, two-storey, semi-detached house will apply a load of about 50 kN/m run to the foundation. A foundation width of 0.5 m would, therefore, apply a stress of 100 kPa to the ground. Such lightweight structures on strip footings typically stress the ground significantly only to depths of 1.5 m to 2.5m. Consequently, it is relatively easy to reproduce the actual stress level and stress distribution with depth in a load test that is simple, cheap and provides direct evidence about the settlement of the foundations. Such tests do have limitations:

- a few load tests may not be representative of the site conditions

- settlement may occur at depth within the ground due to causes other than structural load and the load tests are unlikely to give any indication of this.

The use of a sand-filled rubbish skip as kentledge provides a simple form of test (Charles and Driscoll, 1981). Tests have been carried out by placing the skip on a thin bedding layer of sand, but this can induce some small bedding error and it is preferable to place the skip on a concrete block cast directly on the fill. The use of a concrete block gives greater freedom as the pad can be cast to the required dimensions whereas if the fill is loaded directly by the skip the loaded area has the dimensions of the base of the skip (typically about 1.7 m × 1.7 m). Settlement measurements can be made by precise levelling from bench-marks established away from the test area on ground that is not expected to move significantly during the period of the test.

The actual period over which the test is carried out is a compromise between the following requirements:

- the time-dependent deformation properties of the soil

- the theoretically desirable requirement for a period comparable with the life of the structure

- the practical requirement for early development of the site.

A month is a minimum for the test and it is preferable for tests to be carried out over periods of three to six months, if possible. Settlement is usually best plotted against the logarithm of time elapsed since the load was applied and then results are extrapolated to predict settlement during the lifetime of the structure. An example of the use of skip tests to evaluate the effectiveness of vibro stone columns is provided by the case history presented in Appendix A4.

Where larger and heavier structures will be built on the treated ground, it may be necessary to carry out zone loading tests. In these tests settlement performance is measured over a wider and deeper zone of ground and the results can be used to compare actual performance of the ground with design predictions. The form a zone load test takes will depend on the particular application; it could involve loading a full-scale structural foundation element such as a house raft foundation. In some situations load is applied by temporary earthworks and may be part of a trial embankment study. The cost of this type of testing is high and normally it will only be feasible on large projects.

Figure 5.9 illustrates the way in which a bearing pressure, q, of 50 kPa applied by a footing of width $B =1$ m will only significantly increase the vertical stress, σ_v, in the ground to a depth of about 1.5 m, whereas the same pressure applied by, say, a surcharge of fill with $B =10$ m will significantly increase σ_v to a depth of 7 or 8 m.

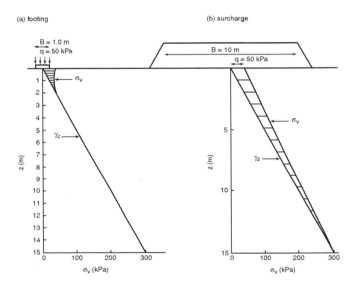

Figure 5.9   *Vertical stress beneath loaded areas*

5.6   MONITORING

It may be advantageous to monitor ground behaviour at several stages of a project:

- prior to ground treatment as part of the site investigation
- during ground treatment as part of the quality control process
- subsequent to ground treatment to confirm satisfactory long-term performance.

The monitoring could include measurement of the following aspects of ground and structural behaviour:

- settlement of the ground
- water levels and pore pressures within the ground
- settlement of the structure.

Surface settlement can be monitored by precise levelling of settlement stations. These stations need not be elaborate, simple steel rods set in concrete may be adequate. It is important to establish stable levelling reference stations and to use sufficiently accurate surveying equipment. Accurate measurements are needed to establish reliably the rate of settlement over a relatively short period. Measurements should be made to an accuracy of at least 1 mm, and corrected for effects such as thermal expansion.

It is also important to measure ground water levels. Simple standpipe piezometers can be sealed into boreholes and the water level can be measured using an electric dipmeter. This will prove satisfactory in many cases but in low permeability ground the response time of the standpipe piezometer to a change in piezometric pressure may be very long. Pneumatic piezometers can be installed in boreholes and will give a much more rapid response. Permeability tests can be carried out in standpipe piezometers.

To reliably establish trends in settlement and ground water levels a minimum period of at least a year will be required and preferably several years. Where a site is being investigated for building development, it will rarely be possible to monitor over such long periods. However even where the period is quite short, it may be advantageous to perform some monitoring.

Post-construction settlement of buildings on fill should be monitored where possible. BRE Digests 343 and 344 (Building Research Establishment, 1989b and 1989c) provide guidance on measuring movements of low-rise buildings. A comprehensive account of appropriate equipment for geotechnical monitoring can be found in Dunnicliff (1988).

6   Provision of ground treatment

The provision of ground treatment and the role of quality management are described with some consideration given to how the effects produced by commonly used treatment methods can improve the load-carrying characteristics of different types of ground. The limitations of ground treatment should be recognised and possible environmental effects anticipated. The performance of the treated ground needs to be adequate for its purpose. In many cases, a principal objective of ground treatment is to increase the stiffness and strength of the ground. In other situations the elimination or mitigation of the potential for collapse compression on inundation or for liquefaction may be key requirements. While examples of unsatisfactory performance are infrequent, possible causes of unsatisfactory performance and processes of deterioration are discussed because much can be learned from situations where problems have arisen.

6.1   TYPES OF TREATMENT

This report is concerned with the properties of treated ground, not with ground treatment methods as such, and therefore does not include detailed descriptions of treatment techniques. Such descriptions can be found in the report from CIRIA, C573 *A guide to ground treatment* (CIRIA, 2002).

This section of the report describes some general matters relevant to ground treatment. In the following three sections treatment techniques are grouped as follows:

- improvement by compaction (Section 7)
- improvement by consolidation (Section 8)
- improvement by stiffening columns (Section 9).

In chapters 7, 8 and 9, the most common forms of treatment are briefly described and their applicability and effectiveness are reviewed in the light of monitored performance. These treatment methods are summarised in Table 6.1. The grouping of case histories associated with each treatment method is helpful in understanding the performance of different types of ground treatment, but in many cases more than one form of treatment is used at a particular site as the following example illustrates.

Box 6.1   *Use of vibro stone columns and vibro concrete columns for road embankment (after Cooper and Rose, 1999)*

Both vibro stone columns and vibro concrete columns were used to support a 7 m high embankment for a roundabout and approach to a bridge abutment at a site on the bank of the River Avon near Bristol where there are 7.5 m deep alluvial deposits. Stone columns were installed at between 2.4 m and 1.5 m centres in the soft clay soils. Vibro concrete columns provided a transition zone between the area improved by stone columns and the piled bridge abutment. A load transfer platform, comprising a 1 m thick granular blanket with three geogrid layers, transferred embankment loads on to the vibro concrete columns, which were designed for end-bearing on an underlying stiff stratum. The combination of ground treatment methods facilitated the rapid construction and commissioning of a large embankment on soft compressible organic soils. Overall settlements were within design criteria.

Table 6.1   *Ground treatment techniques: applicability and testing methods*

Type of treatment process	Treatment method	Applicability	Testing methods
Compaction	Vibro-compaction	• Sandy soils	• *In situ* tests eg CPT, SPT
	Dynamic	• Fills; sandy soils compaction	• *In situ* tests eg CPT, DP, SPT; induced settlement
	Rapid impact compaction	• Coarse fills; sandy soils	• *In situ* tests eg DP; in-cab monitoring of plate penetration; induced settlement
	Excavation and compaction in thin layers	• Non-engineered fills	• Quality control testing during fill placement
Consolidation	Pre-loading without drains	• Partially saturated fills and high permeability soils	• *In situ* tests eg CPT lowering of ground surface
	Pre-loading with drains	• Low permeability soils	• Settlement and pore pressure monitoring
	Lowering ground water table	• Dependent on soil permeability	• Piezometric level and settlement monitoring
	Vacuum pre-loading	• Soft saturated clays with high water table	• Settlement and pore pressure monitoring
	Electro-osmosis	• Silts and silty sands	
Stiffening columns	Vibro stone columns	• Fills; other compressible soils	• Load test
	Dynamic replacement	• Saturated, fine soils	• Pore pressure monitoring
	Stabilised soil columns	• Soft clays and silts	• Column penetrometer; load test
	Vibro concrete columns	• Compressible soils overlying stable stratum	• Load test

Section 3.3.2 explained that the wetting of a partially saturated soil may cause collapse compression and that, if this happens after construction has taken place on the ground, serious problems can result. It might be thought that inundation prior to construction could form a ground treatment technique for increasing the density of partially saturated soils. In practice, it is of limited applicability:

- it is only relevant to certain types of partially saturated soils, mainly fills that are in a loose and/or dry condition

- it is only beneficial where a uniform treatment can be achieved (eg by a rising ground water table); inundation from the surface may produce a variable and unsatisfactory treatment in some types of ground

- it is difficult to control as movements caused by inundation may continue for some time after the addition of water to the ground has ceased; in such situations it is not suitable if construction has to take place immediately following treatment.

6.2 ROLE OF GROUND TREATMENT

Where a site investigation has shown that unacceptable movements may occur over the area of a proposed building or structure, it may be appropriate to improve the load carrying characteristics of the ground by ground treatment, prior to site development. Other management options have been described in Section 3.4.

It has been emphasised in Chapter 3 that, prior to selecting an appropriate ground treatment technique, the problem with the ground should be correctly diagnosed. It is equally important to recognise what can, and what cannot, be achieved by the treatment method. Ground treatment may reduce subsequent movement of the soil or fill, particularly differential movement. However it is unlikely to eliminate movement and so may not be an adequate solution for buildings that are sensitive to small settlements. Typical improvements in properties achieved by the various treatment methods are described in chapters 7, 8 and 9, together with information about subsequent ground movements.

6.3 BASIS OF DESIGN

Different treatment methods in different ground conditions with different types of development in view require appropriate design methods. The design of ground treatment is a large subject and requires specialist knowledge, but some helpful general guidance is available for the design of particular treatment methods, such as:

- vibro stone columns – Priebe (1995)
- vertical drain installations – Hansbo (1993); Holtz *et al* (1991).

While it is not the purpose of this report to describe detailed design methods, it is important to identify some basic principles. The design philosophy should be clearly stated including:

- identification of the specific technical objectives of the treatment
- description of the manner in which the treatment will improve ground conditions and in which deficiencies will be remedied
- quantification of the target performance.

Close liaison between the geotechnical specialist and the structural engineer should aim for compatibility between the required performance of the treated ground and the structural requirements. The objectives of the treatment should be achievable for design loadings that are realistic.

The observational method can be used to advantage with many forms of ground treatment, see CIRIA Report 185 (Nicholson *et al*, 1999). This type of approach requires adequate monitoring and the identification of contingency measures, which can be implemented if the observed performance is not adequate. Table 6.2 gives a simplified view of the way in which the observational method can be applied with vibro densification works.

Table 6.2   *Observational method applied to vibro-compaction (after Raison, 1996)*

Preliminary design	• based on site investigations • moderately conservative conditions • worst case behaviour
Observations	• initial trials • ongoing measurements
Trigger values	• CPT q_c
Contingency measure	• closer centres

6.4   SPECIFICATION AND SITE CONTROL

While the details of the particular contractual arrangements under which ground treatment is carried out are outside the scope of this report, their significance should not be overlooked. An appropriate specification and adequate supervision are vital aspects for successful quality management of ground treatment. The responsibilities and liabilities of the various parties should be clearly understood by all concerned. There has been a rapid development of quality control and quality assurance procedures since the 1970s and there is considerable scope for a more explicit adoption of quality management in ground treatment. Nevertheless, it will still be necessary to make some connection between the parameters that can be controlled and monitored during and immediately subsequent to treatment and the long-term performance of the treated ground. This will be facilitated by laboratory tests and field trials, with associated risk assessments.

The specification should set out the design requirements, be easily understood by the parties to the contract, be practicable, be capable of enforcement and not be unnecessarily costly or time consuming in its application. It should also be capable of being monitored by an effective form of quality assurance procedure with due regard to safety as required by the Construction (Design and Management) regulations (Health and Safety Commission, 1994)

Three approaches to a specification for ground treatment may be identified:

- method specification
- end-product specification
- performance specification.

Care should be taken to avoid creating hybrid specifications that detail procedures and an end-product that are mutually inconsistent.

6.4.1   Method specification

In a method specification, the procedure for the work is specified and site control involves inspection of the works to check compliance with the specified method. The form and detail with which the method is described will depend on the type of treatment being used. For example, with dynamic compaction the total energy input could be specified or more detail could be specified in terms of dimensions of the weights, heights of drop and spacing of compaction points.

A method specification can provide a simple basis for a contract and for a comparison of quotations for the work. However, it does not lend itself to progressive modifications of the treatment process as the work progresses and the treatment itself reveals more about the true ground conditions.

6.4.2 End-product specification

In an end-product specification, a required value for some property or properties of the treated ground is specified. This can be in terms of a parameter measured in an *in situ* test such as the cone penetration test. *In situ* tests that are used for quality control may need to be carried out as soon as a phase of the treatment has been completed, to check that the specified parameter value has been or will be achieved. The results of these tests may not represent the eventual condition of the ground, ie the end-product actually reached when transient effects such as high induced pore pressures are no longer present.

6.4.3 Performance specification

In a performance specification, some quantified aspect of the behaviour of the treated ground related to its performance has to be achieved. This might be in terms of a maximum permissible settlement over a specified period following the completion of treatment or a required result from a loading test. With this approach the specification is directly related to one or more aspects of the structure's performance requirements. It is important that the criterion of ground movement, usually settlement, which must not be exceeded, is realistic both in terms of what it is possible to achieve by ground treatment and what the structure can actually tolerate.

A performance specification may seem attractive to a client because it appears to link to what is required by placing responsibility for such performance on the contractor, which seems contractually clear cut. However, post-construction ground movement is usually the critical criterion and while compliance with such a post-construction performance requirement may be easy to check, it may be difficult to obtain any adequate redress where non-compliance with the specification is established at a late stage. Compliance tested by even a large load test may only demonstrate that the upper layers of ground are adequately treated and may say nothing about the performance characteristics of the underlying ground.

6.5 ENVIRONMENTAL EFFECTS

In addition to improving the load-carrying properties of the ground, a ground treatment technique may have other incidental effects. These can be termed environmental effects and they could have a major bearing on the adoption or rejection of a particular ground treatment technique.

6.5.1 During treatment

Environmental effects during treatment could involve noise, vibration and dust. Operatives, adjacent buildings and their occupants could be affected.

A requirement to move spoil off the site can have severe impacts, both in terms of vehicle journeys on the public highway and the transportation and disposal of contaminants from brownfield sites. A treatment method that eliminates the need for off-site disposal, therefore, has considerable merits.

The provision of water and removal of effluent could be unacceptable. For a treatment method such as rapid impact compaction, noise could be a prohibitive factor. Dynamic compaction can transmit vibrations, which could damage sensitive structures and the risk of flying debris will restrict the acceptable proximity of treatment to other buildings and site activities. Vibro techniques also transmit vibrations but the effect is much more localised. Large earthmoving operations can result in severe dust problems in dry weather.

6.5.2 Subsequent to treatment

Treatment processes that involve the injection of grout, the use of materials such as bitumen and lime, or the installation of drainage could affect aquifers. Vibrated stone columns may allow the upward passage of harmful gases in contaminated or degrading fills into buildings or the surrounding environment, or the percolation of near-surface fluid contaminants down into aquifers, unless specific measures are taken to inhibit such movement.

6.6 UNSATISFACTORY PERFORMANCE

There are very few published reports of unsatisfactory performance for sites that have been subjected to ground treatment, but it is not easy to draw firm conclusions from this absence of data because:

- unsatisfactory performance may have occurred but has not been reported
- the ground might have performed satisfactorily without treatment.

Adequate evaluation of ground treatment requires well documented case histories where field monitoring has been undertaken. Ground treatment has been used on a very large number of sites within the United Kingdom, and, of course, worldwide the number of sites is much greater. Consequently, it might be assumed that there is a correspondingly large amount of performance data. Unfortunately, this is not the case. Case histories provide the necessary background information for realistic predictions of the performance of treated ground. Without such a basis, there is the danger that unrealistic requirements and over-optimistic predictions of performance will be made. Monitored field performance illustrates what can be achieved by ground treatment in terms of the following:

- changes to the overall ground mass properties
- depths of effectiveness achieved by various treatments
- limitation of post-construction settlements.

The critical aspect of performance in many applications of treated ground is post-construction ground movements, that is, those movements that occur after completion of the structure. Although information on long-term performance is scarce, data are available for a number of case histories, which are of great value. Long-term movements may present particular problems in the more cohesive soils and soils with a significant biodegradable content; for coarse soils these movements should be small.

Cases of unsatisfactory performance are not very common and it would be misleading to give them undue prominence. Nevertheless, there is much to be learned from situations where problems have arisen. The definition of unsatisfactory performance needs to be clarified. It might seem simplest to restrict the term to the failure to meet a performance requirement specified in the contract. However, this definition has its limitations as the following situations illustrate:

- A performance requirement specified in the contract is not met, but in practice the performance of the treated ground and associated structure is quite adequate for purpose in terms of visual acceptability, serviceability limit states and ultimate limit states (see Box 6.2).

- All the performance requirements specified in the contract are met, but in practice the performance of the treated ground and associated structure is not adequate for purpose or has unacceptable environmental effects.

Box 6.2   *Case history of factory at Warrington (after Wilde and Crook, 1991)*

Some of the difficulties of defining unsatisfactory performance are illustrated by the case history of a factory at Warrington built on alluvial soils that had been treated with vibro stone columns. During the six years following construction of the factory, some 120 mm of settlement were measured as shown in Figure 6.1, with a maximum differential settlement of 100 mm along the length of the structure. To achieve the required floor level, fill had been placed with a maximum depth of 1.5 m coincident with the maximum 10 m thickness of the alluvial deposit. The settlement was attributable to the weight of the fill, not the lightweight structure. In many circumstances differential movements of the magnitude measured at this factory would be outwith the performance specification and considered to constitute unacceptable performance. However, the building apparently suffered no ill effects.

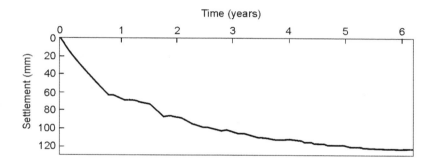

Figure 6.1   *Settlement of factory on vibro stone columns (after Wilde and Crook, 1991)*

Another aspect of unsatisfactory performance is related to unrealistic expectations of what can be achieved by ground treatment. Unsatisfactory performance of the treated ground can result from inadequacies in:

- assessment of required structural performance
- diagnosis of ground problem
- choice of treatment
- design of treatment
- execution of treatment
- appreciation of potential for long-term deterioration of treated ground.

Some of these shortcomings are illustrated by the case histories in Boxes 6.3 and 6.4.

Box 6.3   *Problems at a housing development in West Midlands (after Thompson, 1998)*

The 10 m deep opencast backfill was composed of lumps of stiff clay and highly weathered mudstone. Vibro stone columns were installed beneath the house foundations. Shortly after the development had been completed, some of the properties began to suffer total and differential settlement. The houses were built on reinforced concrete raft foundations and experienced little cracking and damage. Nevertheless, over the 9 m width of a detached house, tilts reached 300 mm. The costly remedial works involved the installation of piles and jacking the houses back to level. Thompson (1998) concluded that the vibro stone columns would have provided a means of introducing water into the backfill and encouraging collapse compression. This case provides a warning of the costs that may be incurred if the condition of the site is not adequately understood at an early stage.

Box 6.4   *Unsatisfactory performance of service station near Peterborough (after Clark, 1998)*

In 1963, a service station was built near Peterborough on fill placed on a natural slope. In 1984, following the placement of more fill to extend the working area, substantial ground movements occurred and buildings cracked. The site was decommissioned and sold. In 1987/1988 drainage measures were undertaken, vibro stone columns were installed in the fill and the face of the fill was regraded. A second service station was then constructed, which also suffered structural damage; displacements were up to 50 mm horizontal and 40 mm vertical. There had been no site investigation prior to construction of the first service station and inadequate investigations following damage to the first structure concluded that movements along the interface between the fill and the natural ground had been caused by over-tipping of the fill material at a steep angle. The true nature of the problem was not identified. Following the movement of the second service station, a desk study and ground investigation revealed that the fill had been placed on a pre-existing landslip, which had been reactivated as a result. Treatment by vibro stone columns had been undertaken despite advice from some of the ground treatment contractors who were invited to tender, who stated that treatment was not appropriate and that a stability analysis should be undertaken.

7 Improvement by compaction

> The *in situ* densification of soils by compaction methods such as vibro-compaction and dynamic compaction is considered in this section of the report. The increase in density of fills and natural sandy soils produced by compaction will generally lead to improved soil behaviour. Case histories of monitored field performance are used to illustrate what can be achieved by ground treatment in terms of improvements in soil properties, the depths of effectiveness of the treatment methods that are applied at the ground surface, and the limitation of post-construction settlement.

7.1 APPLICABILITY OF COMPACTION METHODS

The term *compaction* is used to describe processes in which densification of the ground is achieved by some mechanical means such as rolling, ramming or vibration. Where the major deficiency of the ground is related to its loose state, *in situ* compaction may be the most appropriate type of treatment. Soils are not elastic materials and densification produced by compaction is unlikely to be reversed. Consequently most soils that have been compacted will remain in that dense state.

Two types of ground where this mode of treatment is particularly effective are loose sandy soils and loose heterogeneous fills.

Loose sandy soils usually respond best to vibratory methods. Vibratory rollers will only densify the soil to a very limited depth, so this approach requires excavation of the deposit and compaction in thin layers. Generally it will be preferable to use vibro-compaction in which a large vibrating poker compacts the ground *in situ*. The spacing of the treatment points is a critical element in the design of vibro-compaction.

Both dynamic compaction and vibro-compaction are capable of achieving significant densification of loose, granular deposits. Field investigations indicate that the compressibility of the fill can be greatly reduced, typically to as little as 50 per cent of its original value. Compaction can reduce or remove liquefaction potential. This can be confirmed from the results of post-treatment *in situ* tests. Vibro-compaction and dynamic compaction of sands are some of the simplest forms of treatment to monitor (see CIRIA C573, 2002). *In situ* test measurements can be correlated with density index and hence used to characterise shear strength and compressibility.

The presence of a high water table on site may interfere with ground improvement by compaction. This is the case in fine sands, where compaction can result in an increase in pore water pressure only, with no improvement in the load-carrying properties of the ground.

Loose heterogeneous fills can be compacted by ramming, using techniques such as dynamic compaction and rapid impact compaction. The major limitation is the depth to which surface impact loading can densify the soil. Fills composed of lumps or clods of clay can be compacted by these methods but the effective depth of treatment will be less than with granular fills. Heterogeneous fills can be compacted by excavation and compaction in thin layers. This approach facilitates the removal of unsuitable material.

VIBRO-COMPACTION

The term vibro-compaction is used to describe a deep vibratory technique of improving the properties of granular soils by densification using a large vibrating poker.

Principle

The basic tool is a cylindrical poker, which contains in its bottom section an eccentric weight. Rotation of the weight produces vibrations in a horizontal plane. Radial densification of granular soils is achieved by vibration.

The effectiveness of the treatment will depend on the spacing of the treatment points, the type of equipment and treatment technique, and the responsiveness of the ground to the treatment. The zone of improved soil typically extends from 1.5 m to 4.0 m from the vibrator (Mitchell, 1981)

Methods and equipment

The methods and equipment are similar to those used to install vibro stone columns. Descriptions of the various dry and wet systems of installation are given in Section 9.2. In order to perform vibro-compaction, the vibrating poker is inserted into the ground and penetrates to the required depth by its vibratory action in conjunction with the use of jetting water (Moseley and Priebe, 1993). The vibrating poker is then gradually withdrawn from the ground achieving compaction of the surrounding granular soil by the horizontal vibration forces. A compacted cylinder of soil is formed.

Although granular fills and granular natural soils treated by vibro should benefit from radial densification, almost all vibro applications in the United Kingdom also involve the addition of stone to form stone columns. The addition of stone enhances densification and, by reinforcement, provides further security against the possibility of localised high fines content within the fill or natural soil. The use of stone columns to stiffen the ground is described in Section 9.2.

Testing

The vibro-compaction method of ground treatment can be readily tested by *in situ* tests such as the CPT. Measurements made before and after treatment provide a good indication of the effectiveness of the treatment. *In situ* test measurements can be correlated with density index and hence used to characterise shear strength, compressibility and liquefaction potential.

Applications

The range of particle size, which is generally regarded as being suitable for treatment, is shown in Figure 7.1. It has generally been considered that the method is unlikely to be effective when the fines content (silt and clay size particles) of the ground exceeds about 15 to 20 per cent, but Slocombe *et al* (2000) have shown that the development of new vibrators and modified construction techniques has enabled sands with significantly higher fines content to be treated.

While much vibro ground treatment is carried out only to relatively shallow depths, Degen (1997) has described the vibro-compaction of a 56 m deep sand fill in a German lignite mining area. Depth dependent criteria were defined in terms of CPT results; the required q_c was 5 MPa for depths of less than 5 m increasing to 15 MPa for depths greater than 15 m.

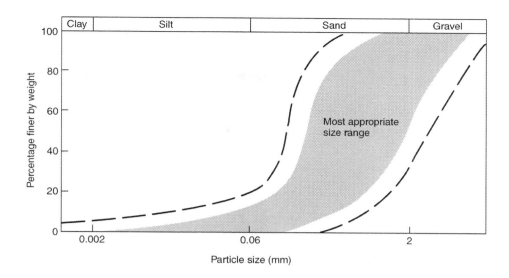

Figure 7.1 *Range of particle sizes for which vibro-compaction should be effective (after Mitchell, 1981)*

Performance

At Belawan Port in Sumatra, Johnson *et al* (1983) reported $I_D = 0.7$ in sand fill with compaction points at 3.5 m centres. $I_D = 0.6$ was required to minimise the risk of liquefaction, increase shear strength and reduce differential settlement.

Some 70×106 m^3 of dredged sand fill was used in the reclamation works for Chek Lap Kok airport in Hong Kong (Covil *et al*, 1997). Below the water table, I_D was in the range 0.2 to 0.4. Vibro-compaction was carried out to densify the sand and produce more uniform conditions. The grid spacing was related to the type of rig with light compaction carried out on a 4.0 m triangular grid and heavy compaction on a 3.5 m grid. A minimum q_c of 8 MPa was specified for light compaction and 15 MPa for heavy.

Case histories of the performance of vibro-compaction at sites in the United Kingdom are given in Boxes 7.1 and 7.2

Box 7.1 *Vibro-compaction for housing development (after Watts and Charles, 1991)*

An alluvial sand was treated for a housing development at Wythenshawe. Stone columns were installed using the wet method (Section 9.2). Values of CPT q_c and DMT p_0 and p_1 parameters increased by up to 100% within 0.5 m of columns post-treatment, but with little improvement at distances greater than 2 m from the columns. The long-term performance of the houses built on the treated ground was monitored by precise levelling of points installed at DPC level as shown in Figure 7.2. Figure 7.3 shows the very small settlement of the blocks of low-rise houses. The greatest settlement occurred where a 0.35 m thick peat layer was present at shallow depth below the strip foundations. Settlements were less than one half of those computed for untreated ground.

Figure 7.2 *Monitoring settlement of houses built on vibro stone columns (courtesy of Building Research Establishment Ltd)*

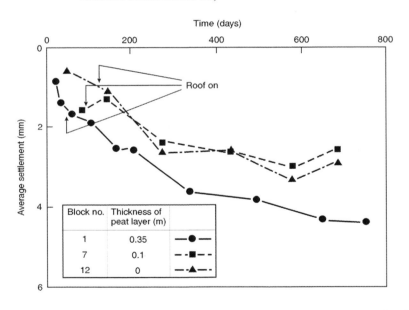

Figure 7.3 *Settlement of houses on sand with peat layer (after Watts and Charles, 1991)*

Box 7.2 *Vibro-compaction for bridge foundations (after O'Brien, 1997)*

A bridge was required to carry a new road over a railway line at a site in Flintshire where there was fine, loose, estuarine sand. A preliminary trial of vibro-compaction was carried out and a spacing of 2.2 m was chosen for the main works. Figure 7.4 shows the improvement measured in CPT q_c values at the southern bridge abutment. Cone-pressuremeter testing indicated that the density index had increased from 0.4–0.45 to 0.7–0.85. The short-term settlement measured under a bearing pressure of 300 kPa was 5 mm to10 mm; the increased q_c values indicate that settlement had been reduced to 30 per cent of the settlement predicted without treatment.

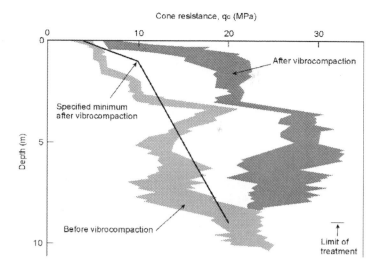

Figure 7.4 *Vibro-compaction of sand fill (after O'Brien, 1997)*

At some sites both vibro-compaction and vibro stone columns are used. At a bridge approach in Illinois liquefiable deposits comprising 6 m of soft silt overlying 21 m of loose sand were treated by installing 0.9 m diameter vibro stone columns in the soft silt and using vibro-compaction to densify the top 11 m of the underlying loose sand (Zdankiewicz and Wahab, 1999). Treatment points were at 2.4 m centres.

Key features

1   Vibro-compaction can achieve significant densification of loose, coarse soils. It is unlikely to be effective when the fines content of the ground exceeds about 15 to 20 per cent.

2   Vibro-compaction is associated with low levels of noise and vibration.

3   Vibro-compaction is of limited success on sites containing buried obstructions.

4   The spacing of the compaction points is a critical element in the design of vibro-compaction. The extent of effective radial densification is a function of soil properties such as particle size distribution and required density index, and the characteristics of the vibrating poker. Treatment points are typically at 2 m to 3 m centres.

5   On a housing development, treatment can be provided at the location of strip foundations and it is not necessary to treat the whole site. Therefore this method can be cost-effective for small sites.

6   *In situ* test measurements, such as the CPT, can be made before and after treatment to monitor the effectiveness of the treatment. The measurements give an indication of increased density; and published correlations between *in situ* measurements and density index make it possible to assess shear strength and compressibility.

7   Field investigations indicate that the compressibility of loose sand can be much reduced, sometimes to 50 per cent of its original value. Liquefaction potential can be greatly reduced or eliminated.

8   Post-construction ground movements are not usually a problem with sandy soils and this is confirmed by monitored performance.

DYNAMIC COMPACTION

The repeated dropping of a weight on to the ground surface is one of the most basic methods of ground compaction and this type of soil treatment has a long history (Slocombe, 1993). The Menard technique of "dynamic consolidation" using high energy impacts was introduced into the United Kingdom in the 1970s (Menard and Broise, 1975; Anon, 1979).

Principle

Deep compaction is effected by repeated impacts of a heavy weight on the ground surface. This method of compaction is analogous to the standard laboratory Proctor compaction test but the energy per blow is typically 1×10^5 times as large.

Methods and equipment

A crane is required of sufficient capacity to drop the specified weight from the required drop height. Originally weights of up to 15 tonnes were dropped from heights of up to 20 m. More recently weights of 8 to 15 tonnes have been dropped from heights of up to 15 m to achieve treatment depths of about 6 m; the greater the energy input, the greater the depth of treatment. Figure 7.5 shows dynamic compaction at a filled site.

Testing

Where a loose sand is compacted by dynamic compaction, the effectiveness of the treatment can be readily evaluated by *in situ* tests such as the CPT. Measurements made before and after treatment provide a good indication of the effectiveness of the treatment in improving properties and the depth to which improvement has been achieved. *In situ* test measurements can be correlated with density index and hence used to characterise shear strength, compressibility and liquefaction potential.

With clay fills and heterogeneous waste fills, *in situ* testing is more difficult to interpret. The presence of obstructions may make the CPT impractical and the SPT and DP may be more suitable. An example of the use of Menard pressuremeter tests in pre-drilled boreholes in a waste fill is given in Appendix A.1.

Figure 7.5   *Dynamic compaction of coarse fill (courtesy of Building Research Establishment Ltd)*

Applications

The major use of the treatment method in the United Kingdom has been the deep compaction of loose, partially saturated fills (Charles *et al*, 1981). The average surface settlement induced by treatment gives some indication of the effectiveness of the treatment. BRE research (Charles, 1993) has investigated the depth and degree to which fill can be compacted, and the magnitude of movements that will occur in the long-term after ground treatment.

Potential problems include vibration damage to adjacent structures and flying debris. These hazards, together with relatively high plant mobilisation costs generally restrict the use of this technique to larger, more open sites.

Performance – sand

In situ measurements made before and after dynamic compaction of hydraulic sand fills at three sites in South East Asia are summarised in Figure 7.6.

- At Changi airport in Singapore the reclamation project involved dredging over 36×10^6 m^3 of sand. The sand was pumped through a 4 km long pipeline before being discharged in the area of reclamation (Radhakrishnan *et al*, 1983).
- At Pulau Ayer Merbau, also in Singapore, the ground was treated for the construction of a warehouse.
- At Ashuganj in Bangladesh the ground was treated for a fertiliser factory.

Table 7.1 presents the density index values inferred from SPT, CPT and PMT results by Ramaswamy and Yong (1983) using published correlations. The significant differences between the values deduced from the three test methods could be related to deficiencies in the correlations, which have been used as well as to scatter in test results associated with lack of homogeneity in the sand fill deposits.

Typically, dynamic compaction increased the density index from 0.5 to 0.8. Dynamic compaction was effective at Changi despite the presence of a fines content in the hydraulic fill as large as 30 per cent (Choa *et al*, 1979).

Table 7.1   *Density index of sand fill deduced from* in situ *tests (after Ramaswamy and Yong, 1983)*

Location	Condition	Density index		
		SPT	CPT	PMT
Changi	before compaction	0.5	0.6	0.6
	after compaction	> 0.9	0.75	0.85
Pulau Ayer Merbau	before compaction	0.3	0.35	0.25
	after compaction	0.65	0.6	0.55
Ashuganj	before compaction	> 0.9	0.5	0.4
	after compaction	> 0.9	> 0.9	0.6

Performance – clay and heterogeneous fills

The major use in the United Kingdom has been the deep compaction of loose, partially saturated fills. BRE research has investigated the depth and degree to which fill can be compacted and the magnitude of movements that will occur in the long term after ground treatment (Charles *et al*, 1981; Charles, 1993).

Dynamic compaction has been used on a wide variety of fill materials including miscellaneous waste fills and clay fills. Measuring the improvement in the properties of these fills is not as simple as it is with sandy soils. However, the depth by which the ground surface is lowered gives an indication of the improvement in ground properties. Settlement induced by dynamic compaction has been monitored at depth at a number of filled sites and an example is presented in Box 7.3.

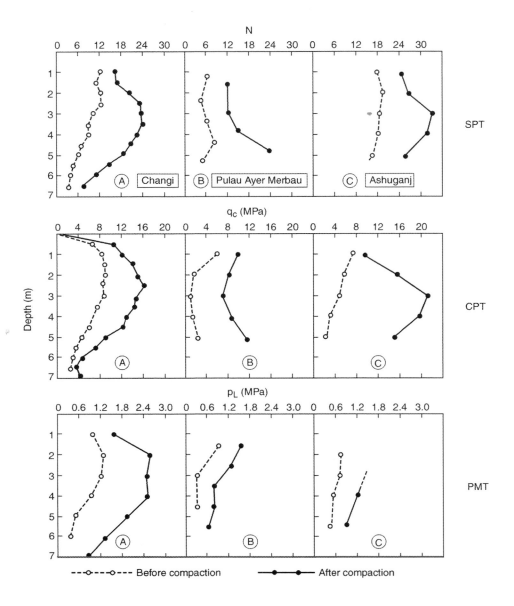

Figure 7.6 *Dynamic compaction of three sand fills (after Ramaswamy and Yong, 1983)*

Box 7.3 *Depth of effectiveness of dynamic compaction of clay fill (after Charles et al, 1978)*

An experimental site on opencast ironstone mining backfill at Snatchill, Corby was divided into four 50 m x 50 m square areas. Three ground treatment methods were used on the predominantly clay fill; dynamic compaction, pre-loading with a 9 m high surcharge of fill and pre-inundation from 1 m deep trenches. The fourth area was untreated. Dynamic compaction involved dropping a 15 tonne weight from heights of up to 20 m. Compaction was carried out in several stages and the total energy input was 280 tonne-m/m². The settlement induced by dynamic compaction at depth within the clay fill is shown in Figure 7.7. The measurements of settlement at depth were made at five settlement gauges and the mean surface settlement was measured by precise levelling. Compaction was effective to a depth of about 6 m.

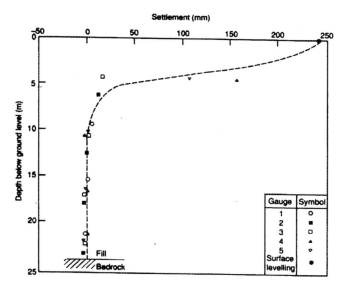

Figure 7.7   *Dynamic compaction of clay fill (after Charles et al, 1978)*

Long-term settlement can be a particular problem for clay fills and for fills containing biodegradable material. Dynamic compaction should improve the properties of such fills in most cases, but is unlikely to eliminate long-term movement. Examples are presented in Boxes 7.4 and 7.5.

Box 7.4   *Long-term settlement following dynamic compaction of clay fill (after Charles, 1993)*

Houses were built on each area at the Snatchill experimental site at Corby and the long-term settlements measured in the ground treated by dynamic compaction are plotted on Figure 7.8. The figure shows the settlement of the houses and settlement of treated ground that does not have the weight of a building on it. The rates of settlement are similar indicating that the settlement of the houses is not principally caused by the weight of the houses.

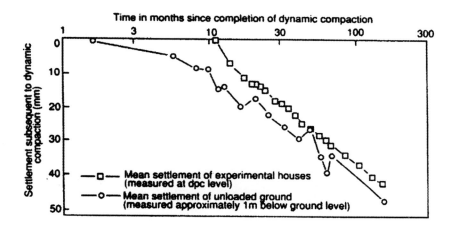

Figure 7.8   *Long-term settlement of clay fill treated by dynamic compaction (after Charles, 1993)*

Box 7.5   *Long-term settlement following dynamic compaction of old domestic refuse (after Charles, 1993)*

A dual carriageway road with interchanges and slip roads was built across a 6 m depth of old domestic refuse at Redditch. The refuse was treated with dynamic compaction and then a 3 m high embankment was built for a slip road. The surface of the refuse settled immediately by about 20 mm. The continuing movements measured at two locations at the surface of the refuse are shown in Figure 7.9. A logarithmic time scale is used based on the time that has elapsed since the refuse was deposited. Dynamic compaction produced an average enforced settlement of the ground surface of 0.5 m representing a volume reduction of the refuse of about 10 per cent. The constrained modulus of the refuse was increased to more than three times its original value. However, treatment has not eliminated long-term settlement. The movements that have occurred are not a problem for a road embankment, but would have been significant if buildings had been founded on the treated ground.

A further case history of the use of dynamic compaction is given in Appendix A1. This describes the treatment of a waste fill for a warehouse development at Cwmbran.

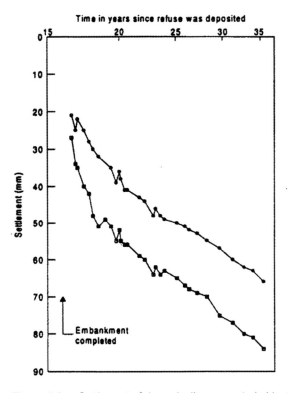

Figure 7.9   *Settlement of dynamically compacted old refuse under embankment (after Watts and Charles, 1999)*

Performance – depth of effectiveness

Where ground treatment applies some form of loading to the ground surface, it is important to know the depth to which the treatment is effective. This applies to both surcharge loading and dynamic compaction. Menard and Broise (1975) suggested that if a weight of W tonnes was dropped from a height of H metres the thickness of layer, z_e metres, which would be compacted was such that $z_e < (WH)^{0.5}$. Field data have confirmed the validity of this inequality, but in practical terms such an inequality is of limited use. A relationship of the following form is of more assistance:

$$z_e = k \, (WH)^{0.5} \tag{7.1}$$

CIRIA C572

A large quantity of field data, summarised by Mayne *et al* (1984), indicated that the coefficient k is generally in the range 0.4 to 0.8. A clay fill is likely to have a value of k towards the lower end of the above range (Charles *et al*, 1981) whereas granular soils and fills will have values towards the upper end of the range (Chow *et al*, 2000). The relationship provides a crude indicator of the thickness of the layer of soil that will be compacted and is helpful in the early stages of assessing the utility of dynamic compaction at a particular site. However, the field data on which the relationship is based may have used different criteria for depth of effectiveness, and the relationship does not incorporate all the significant parameters. The total energy input affects the depth of effectiveness of the treatment and this is particularly important for rapid impact compaction where there will be a much larger number of impacts. The data in Table 7.2 illustrates the effect of dynamic compaction and rapid impact compaction on various types of filled ground.

Table 7.2 *Field compaction of fills (after Charles, 1993)*

Method	Fill	W (tonne)	H (m)	A (m²)	n	E_a (tonne.m/m²)	S (m)	z_t (m)	z_e (m)	k
DC	Stiff clay	15	20	4	3.7	280	0.24	24	6	0.35
DC	Old refuse	15	20	4	3.5	260	0.5	6		
DC	Old refuse	14	14	4	5.3	260	0.58	6.5	5.5	0.39
DC	Old refuse	15	20	4	2.9	220	0.5	8		
RIC	Building waste	7	1	1.8	39	150	0.3	6.5	4	1.5

Notes:

W = weight of tamper, H = height of fall,

A = area in contact with fill on impact, n = number of impacts at any point,

E_a = total energy input per unit area, s = induced surface settlement,

z_t = depth of fill, z_e = depth of effectiveness of treatment,

k = coefficient in equation: $z_e = k(WH)^{0.5}$ RIC = rapid impact compaction.

DC = dynamic compaction,

Key features

1 Dynamic compaction can be effective on natural coarse soils and a wide range of loose, partially saturated fills.

2 The density index of sand fill can be increased typically from 0.5 to 0.8 and treatment can be effective despite the presence of a fines content as high as 30 per cent.

3 The energy per blow has a major influence on the depth of effectiveness. A relationship of the form $z_e = k (WH)^{0.5}$, with k generally in the range 0.4 to 0.8, gives an indication of the depth of soil that will be compacted. A clay fill is likely to have a value of k towards the lower end of the above range whereas sandy soils will have values towards the upper end of the range.

4 Dynamic compaction is unlikely to eliminate long-term settlement in clay fills and fills containing biodegradable material.

5 There are high mobilisation costs associated with the large crane required to drop the weight and, therefore, the method is only likely to be economic on large sites. A granular blanket is likely to be required to support the crane.

6 Flying debris constitutes a hazard for personnel, vehicles or structures close to the impact point. Some shielding of vulnerable targets may be required.

7 Dynamic compaction produces high levels of vibration and noise.

RAPID IMPACT COMPACTION

The technique was developed for the rapid repair of explosion damage to military airfield runways.

Principle

As in dynamic compaction, the ground is compacted by impacts on the ground surface. The main difference between rapid impact compaction and dynamic compaction is that in rapid impact compaction the plate that transmits the energy to the ground remains in contact with the ground. The energy per blow is small compared with dynamic compaction but the rate and number of blows is considerably greater, which can result in a greater total energy input per unit area of treatment.

Methods and equipment

A modified hydraulic piling hammer acts on a 1.5 m diameter articulating compacting foot. The plant is much smaller than that required for dynamic compaction and mobilisation costs are lower. The rapid impact compaction of granular fill is shown in Figure 7.10.

Testing

The method lends itself to in-cab monitoring systems in which the successive amounts by which the plate has been driven into the ground are measured after each impact.

The description of the use of *in situ* test methods given in Section 7.3 on dynamic compaction is equally applicable to rapid impact compaction.

Applications

Rapid impact compaction is a promising technique for the improvement of miscellaneous fills of a generally granular nature, including building and demolition waste and ash fills, and sand fills for depths of about 4 m, although substantially greater depths have been achieved in favourable ground conditions.

Although the method is comparable in fundamental principle to dynamic compaction, a major use is likely to be as an alternative to vibro on small sites, where shallow fills are present, which are being redeveloped for low-rise construction. However, vibration and noise levels could restrict its use at some locations.

Performance

A number of case histories have been described by Watts and Charles (1993). Detailed investigations were carried out at four sites, including measurements of settlement induced by treatment at different depths within the fill and measurements of dynamic probing resistance before and after treatment. The results confirmed that miscellaneous fills of a generally granular nature could be improved to a depth of 4 m.

Key features

1   Rapid impact compaction can improve miscellaneous fills of a generally granular nature, building and demolition waste and ash fills and sand fills for depths of about 4 m.

2   The method lends itself to in-cab monitoring and recording techniques.

3   The method is less effective than dynamic compaction in the treatment of clayey fills, particularly at depth.

Figure 7.10 *Rapid impact compaction of granular fill (courtesy of Building Research Establishment Ltd)*

7.5   COMPACTION IN THIN LAYERS

Where treatment methods are implemented from the ground surface, there are limitations on the depth to which the ground can be effectively treated. Excavation and compaction in thin layers provide an alternative approach. Where fill is imported, it may be necessary to comply with waste regulations. In such cases advice from the Environment Agency should be sought.

Principle

A remedial treatment for non-engineered fills is to excavate the fill with deficient load-carrying properties and replace it with an engineered fill placed in thin layers with adequate compaction. The engineered fill may comprise suitable imported material or the original material compacted to a satisfactory state. With the latter approach, zones of unsuitable material can be identified and removed from the site during excavation.

Methods and equipment

The method involves bulk earthmoving, as does pre-loading with a surcharge of fill (sections 8.2 and 8.3). Parsons (1992) has given a comprehensive review of the performance of compaction plant on different types of soil and granular material. The compaction of an opencast backfill is shown in Figure 7.11.

A specification for engineered fills appropriate for the particular type of development should be used. Trenter and Charles (1996) have described an appropriate specification for building developments. A specification for engineered fills for road embankments may not be appropriate for a building development. Charles *et al* (1998) have shown that the commonly adopted 95 per cent relative compaction requirement will not always be adequate, particularly with fine soils.

Figure 7.11 *Compaction in layers of an opencast mining backfill (courtesy of Building Research Establishment Ltd)*

Testing

Where a method specification is adopted, quality control testing will be required to confirm that the specification is being met. Trenter and Charles (1996) recommend suitable rates for density and compaction testing.

Applications

The method is appropriate for non-engineered fills where the principal deficiency is their loose or uncompacted condition. An example of this approach is provided by the ground treatment carried out in the development of a new township at Peterborough.

Box 7.6 *Housing development on engineered clay fill*

> For over 100 years, Lower Oxford clay was excavated at Fletton, Peterborough for brick-making and much of the site was derelict for decades. There were several different landforms including PFA filled pits (Humpheson *et al*, 1991) and ridge-and-furrow formation. When it was decided to redevelop the area as a township, the ridge-and-furrow area required a major earthmoving operation (Patel, 1995). Much of the clay that was excavated and then compacted in thin layers as an engineered fill was considerably wet of optimum moisture content with a low undrained shear strength. A minimum undrained shear strength of 30 kPa was specified for foundation design.

Performance

While there are few published records of the behaviour of fill that has been excavated and recompacted in thin layers for subsequent building development, there are many relevant case histories, which describe the performance of engineered fills in embankment dams and highway embankments. There are also a number of well documented studies of opencast mining backfills placed as engineered fills (Clarke *et al* [eds], 1993)

In some situations it may be appropriate to improve the properties of the recompacted fill by the addition of lime and/or cement. Tonks *et al* (1998) have described this approach at a 32 000 m² distribution centre. A schematic cross-section is shown in Figure 7.12. The lime-treated clay soil was compacted in 350 mm layers with 12 passes of a 3 tonne/m vibrating smooth wheel roller. The 350 mm finishing layer was treated with lime and cement and topped with 100 mm of compacted granular fill.

Key features

1 The method is particularly appropriate for non-engineered fills where the major deficiencies are associated with heterogeneity of the fill and looseness resulting from lack of adequate compaction.

2 This method is particularly cost effective for large sites and for shallow treatment.

3 The fill should be placed and compacted to a specification appropriate for building developments.

4 Quality management should include an appropriate level of supervision and monitoring.

5 This method is difficult to use if groundwater is present.

6 This method can be difficult to use in contaminated ground.

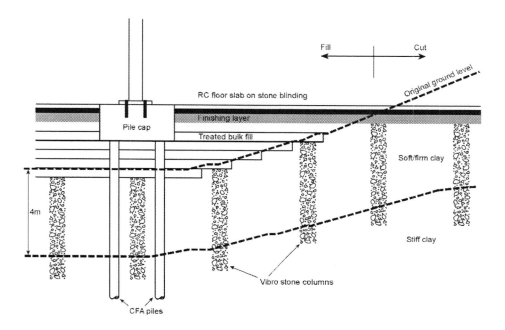

Figure 7.12 *Ground treatment for distribution centre (after Tonks et al, 1998)*

8 Improvement by consolidation

The *in situ* consolidation of soils by ground treatment methods, such as pre-loading and lowering of the ground water table, is considered in this section. Consolidation can be defined as a process in which densification is achieved by an increase in effective stress usually applied by some form of static loading such as a surcharge of fill material. As with compaction, an increase in density will generally lead to improved soil behaviour. Case histories of monitored field performance demonstrate what can be achieved by ground treatment in terms of improvements in soil properties, the depths of effectiveness of various treatments and the limitation of post-construction settlements.

8.1 APPLICABILITY OF CONSOLIDATION METHODS

Consolidation makes the ground stiffer under subsequent applied loads and the higher than expected load-carrying characteristics of many natural soils can be attributed to pre-loading during their geological history. One of the most fundamental methods of ground improvement is temporarily to pre-load the ground prior to construction. The history and application of this type of ground improvement have been reviewed by Stamatopoulos and Kotzias (1985).

Soils are not elastic materials and densification produced by consolidation is unlikely to be reversed. However, there is likely to be some small amount of heave following removal of a temporary surcharge of fill, but this will usually occur immediately following removal of the surcharge and have little effect on subsequent construction.

As would be expected from the fundamental nature of this type of ground treatment, the method can be usefully applied to a wide range of ground conditions including:

- fills
- soft natural clay soils.

The simplest approach is to consolidate using a temporary surcharge of fill. A relatively large area is needed for the process to be practical and the cost will be largely controlled by the haul distance, so a local supply of fill is required. Consolidation of low permeability, soft, natural clay soils will take time and it is usually necessary to install vertical drains to speed consolidation.

Pre-loading can greatly reduce compressibility. Provided that the structural load is significantly smaller than the pre-loading, the treated ground has been effectively over-consolidated and the compressibility will have been reduced to as little as 10 to 20 per cent of its untreated value. The ratio of the ground heave that occurs when the pre-loading is removed to the settlement that had occurred when the ground was pre-loaded, gives a crude indication of the likely improvement in constrained modulus. Both natural soils and fills can be improved by pre-loading with a surcharge of fill. However, there is a major difference in the rate at which consolidation occurs in low permeability, saturated, natural soils and the rate at which it occurs in partially saturated fills. This leads to significant differences in the way that this type of treatment is applied.

8.2 PRE-LOADING WITHOUT INSTALLING DRAINS

Temporary pre-loading with a surcharge of fill can be a very effective means of improving the load-carrying properties of compressible soils prior to building development.

Principle

The properties of fills can be improved by temporarily pre-loading, which over-consolidates the *in situ* fill before construction takes place at the site. With partially saturated fills, compression will generally occur as the surcharge is placed due to reduction in air voids. In these situations there is no need to install vertical drains to speed-up the consolidation process.

Methods and equipment

This type of treatment does not require specialist equipment. Although this should be a considerable advantage, in practice it has meant that there has been little commercial incentive to promote the technique. Consequently, it is sometimes passed over for less appropriate proprietary methods. A common solution to the requirement for a local supply of surcharge fill is to use, say, the top 2 m of fill from a large area of the site. Figure 8.1 shows the pre-loading of opencast mining backfill with a surcharge of fill prior to building low-rise housing on the site.

Figure 8.1 *Pre-loading with surcharge of fill courtesy of Building Research Establishment Ltd)*

The settlement of the ground surface induced by pre-loading gives a valuable indication of the effectiveness of the treatment, but does not indicate the depth to which the treatment has been effective.

Testing

In situ penetration tests can have a useful role in evaluating ground improvement of granular fills. In fills that are composed of clods of clay, ground treatment will reduce the macro-voids, but penetration tests are unlikely to detect this. The overall lowering of the ground surface induced by pre-loading will give an indication of the effectiveness of the treatment.

Applications

In the United Kingdom temporary pre-loading with a surcharge of fill has been used to improve fills prior to building development (Charles *et al*, 1986; Charles, 1993). Where partially saturated fills are pre-loaded, the major effect will be a reduction in the percentage air voids. In contrast to soft saturated clay soils, compression will generally occur immediately load is applied. Even in a clay fill, the consolidation will be largely associated with the volume reduction of the macro-voids between the lumps of clay, and the compression of the fill will occur mainly as the loading is applied.

Performance

A clay fill was pre-loaded at the Snatchill experimental housing site in Corby (Charles *et al*, 1978). Figure 8.2 shows that the 0.5 m settlement induced by the 9 m high surcharge occurred during the three weeks of surcharge fill placement with only very small additional movements during the four weeks that the fill was left in place. Further information is given in Appendix A.2.

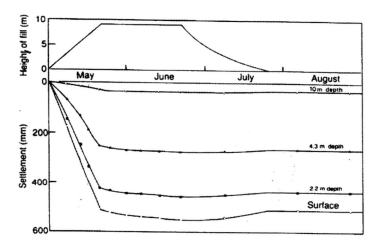

Figure 8.2   *Pre-loading of clay fill (after Charles* et al, *1978)*

Houses were built on the experimental site at Corby. In addition to the pre-loaded area, one area was treated by dynamic compaction (Section 7.3), one area was treated by inundation from 1 m deep surface trenches and the fourth area was untreated. The maximum, minimum and typical long-term settlements of houses on each area are plotted on Figure 8.3. In all areas some movements were continuing after 15 years. Settlement has been smallest in the pre-loaded area.

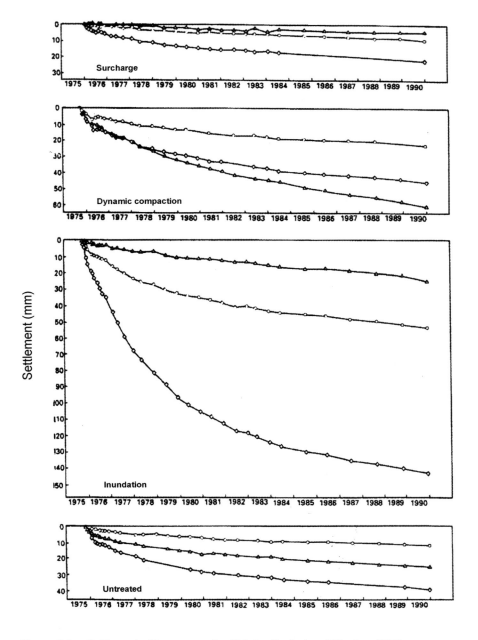

Figure 8.3 *Settlement of houses on clay fill (after Burford and Charles, 1991)*

The behaviour of a 32 m deep opencast mining backfill in Staffordshire has been monitored since 1977 and significant settlements have been measured throughout this period (Charles and Burford, 1987). The fill is predominantly clay with shale fragments. Pre-loading was carried out in a phased earthmoving operation prior to building development. Typically there was a temporary 7 m net increase in depth of fill over an area of 70 m × 50 m. Settlements induced at original ground surface ranged from 190 mm to 230 mm. There was a significant effect to between 6 m and 12 m below original ground level, with some effect to 20 m depth. Most of the settlement took place during the two and a half weeks of mound construction with only small additional movements continuing while the surcharge was left in place for three months. Up to 30 mm of rebound was measured on removal of the surcharge.

It is important to be able to assess the depth, z_e , to which pre-loading with a surcharge of fill is likely to be effective. Table 8.1 lists a number of sites where BRE has carried out field studies.

Table 8.1   *Field studies of pre-loaded fills*

	Fill type	Location	References
1	Clay	Snatchill, Corby	Charles *et al*, 1978; 1986
2	Clay	Millcroft, Corby	Burford, 1991
3	Mudstone and sandstone	Horsley, Northumberland	Charles *et al*, 1977; 1986; 1993
4	Lagoon PFA	Peterborough	Charles *et al*, 1986
5	Old refuse	Liverpool	Charles *et al*, 1986
6	Clay with shale fragments	Staffordshire	unpublished

Figure 8.4 is based on the field studies of pre-loaded fills listed in Table 8.1 and shows the relation between the ratio of the depth of effectiveness to the height of the surcharge, z_e/H, and the ratio of the height of the surcharge to the width of the surcharge, H/B. Since z_e is a function of the ratio of the increment of vertical stress produced by the surcharge to the existing overburden stress, H has been multiplied by a factor, γ_s/γ, where γ_s is the bulk unit weight of the surcharge and γ is the effective unit weight of the loaded ground.

The effective depth of influence of the surcharges is based on the depth at which 90 per cent of the settlement had occurred. Attempts to estimate z_e from simple linear elastic theory tend to over-estimate it when compared with field data and Figure 8.4 shows a relationship derived by Charles (1996), which gives better agreement with the field data. In the range of most practical interest with $0.1 < H/B < 0.5$, z_e/H is likely to be of the order of 1.0 to 2.0.

Although many of the applications of pre-loading without the need to install vertical drains involve the treatment of partially saturated fills, there are occasions where natural soft clay soils may be successfully treated in this manner. The construction of a railway embankment at Irlam has provided a case history of such a pre-loading, which has some unique features and is described in Box 8.1.

Box 8.1   *Polystyrene embankment for railway use (after O'Brien and Anderson, 1999; Anon, 1999)*

An old bridge over a dried-up river bed has been replaced with an embankment at a site in Irlam, where an 8 m deep infilled channel containing very soft, silty clay presented a major problem. Pre-loading with a 4.5 m high embankment of granular fill produced 175 mm of settlement in a nine-month period. The granular fill was then removed. The 14 m high embankment was constructed of expanded polystyrene (EPS) blocks. This ultra-lightweight material had densities varying from 20 to 55 kg/m³ for the five grades of EPS that were used in the embankment. The blocks were arranged in a staggered pattern to avoid continuous vertical joints. The embankment contained two reinforced concrete slabs at intermediate levels and granular fill was placed to form the embankment slopes and protect the EPS. A row of concrete troughs were placed on the top of the embankment.

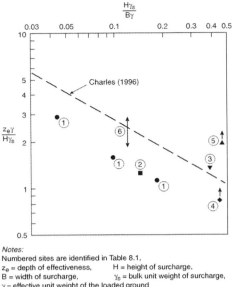

Notes:
Numbered sites are identified in Table 8.1,
z_e = depth of effectiveness,   H = height of surcharge,
B = width of surcharge, γ_s = bulk unit weight of surcharge,
γ = effective unit weight of the loaded ground

Figure 8.4   *Depth of influence of surcharge loading of fills*

Key features

1   Temporary pre-loading with a surcharge of fill can be a very effective means of improving the load-carrying properties of fills prior to building development.

2   Where partially saturated fills are pre-loaded, the major effect will be a reduction in the percentage air voids. Even in a clay fill, the consolidation will be largely associated with the volume reduction of the macro-voids between the lumps of clay and the compression of the fill will occur mainly as the loading is applied.

3   The economic viability of the method is strongly dependent on a local supply of surcharge fill.

8.3   PRE-LOADING WITH INSTALLATION OF DRAINS

The load-carrying properties of over-consolidated clay soils are generally superior to those of normally consolidated clay soils. On a relatively small scale, ground treatment can produce similar effects to those produced by geological processes of deposition and erosion.

Principle

When a natural, soft, saturated clay is loaded, the rate at which the high pore pressures induced by the loading dissipate will be controlled by the low permeability of the soil and consolidation will occur slowly. In many cases it will be necessary to install vertical drains to accelerate the process. A mathematical analysis of the problem using Terzaghi's theory of consolidation was developed by Barron (1948). Temporary pre-loading should reduce the compressibility and increase the strength of a soft clay soil.

Methods and equipment

The rate at which the high pore pressures, induced by the loading, dissipate will be controlled by the low permeability of the soil. Vertical drains can accelerate the consolidation of a saturated clay soil and such drains may be required in order to effectively pre-load the soil in an acceptable period of time. Sand drains, sand wicks and prefabricated band drains have been used.

Band drains are normally installed inside a steel mandrel, which is either pushed or driven into the ground. At the bottom of the mandrel a steel anchor plate is attached to the drain to shut off the bottom of the mandrel and provide a reliable anchoring at the required depth. When the mandrel reaches the required depth, it is pulled out leaving the anchored drain in position. There are many proprietary band drains (Hansbo, 1993), generally consisting of a geosynthetic filter surrounding a plastic core. Band drains typically have a width of 100 mm and a thickness of between 3 mm and 6 mm.

Testing

Monitoring of settlement and pore pressures during pre-loading should provide an adequate understanding of the effectiveness of the ground treatment.

Applications

The first development of prefabricated drains was reported by Kjellman (1948). The efficiency of cardboard band-shaped drains was studied at a soft clay test site at Vasby, north of Stockholm (Chang, 1981). They have been widely used on soft clay sites to speed-up consolidation during pre-loading with a surcharge of fill.

Performance

A test involving the installation of two types of prefabricated band drain, sand drains and an untreated area was carried out at Orebro in central Sweden (Eriksson and Ekstrom, 1983); the settlement monitored in the four areas under a 2.1 m height of fill is shown in Figure 8.5. The drain spacing was 1.4 m on a square grid.

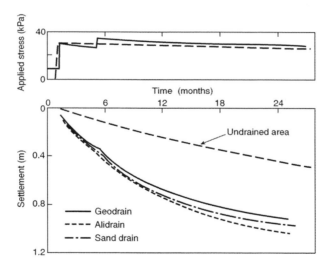

Figure 8.5   *Consolidation of soft clay using vertical drains (after Eriksson and Ekstrom, 1983)*

One limitation of data of this type is that it does not continue for a sufficient length of time to establish the final settlement in the untreated ground. Settlement has occurred more rapidly in the early stages with vertical drains, which could indicate that the drains have successfully speeded-up the consolidation process. However, an alternative interpretation is that installation of the drains has led to increased total settlement.

In an organic soil, secondary compression may be significant. This will limit the effectiveness of vertical drains, which can only increase the rate of primary consolidation.

Settlement measurements were carried out on test embankments on amorphous peat and gyttja at the Antoniny site in Poland. (Gyttja originates from the remains of plants and animals rich in fats and proteins, in contrast to peat, which is formed from the remains of plants rich in carbohydrates.) Geodrains were installed at 1.2 m centres on a square grid. A 4 m high embankment was built in three stages. The settlements monitored during these stages at locations with and without drains are shown in Figure 8.6. It was concluded that the vertical drains did not have a major effect.

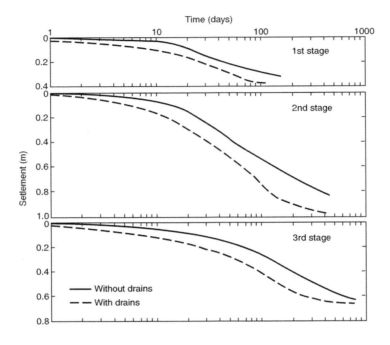

Figure 8.6   *Consolidation of organic soil using vertical drains (after Wolski, 1996).*

At Monnickendam, north of Amsterdam, prefabricated drains were installed at 2 m centres through layers of clay and peat. A 5 m high embankment was built and 1 m of settlement occurred. Two years after drain installation, a trench was cut to form a small tunnel for cyclists. This exposed some of the drains and substantial buckling was observed in a soil layer where compressions of 30 to 35 per cent had occurred (Van Santvoort, 1994). This can significantly reduce the discharge capacity of the drain.

Key features

1   The high pore pressures induced by pre-loading soft saturated clay will be controlled by the low permeability of the soil and consolidation will occur slowly. In many cases it will be necessary to install vertical drains to accelerate the process.

2   Prefabricated band drains are now more commonly used than sand drains.

3   Where large compressions occur, band drains may buckle reducing the discharge capacity.

4   Vertical drains should increase the rate of primary consolidation, but where there is significant secondary compression, as with organic soils, their effectiveness will be limited.

8.4 GROUND WATER TABLE LOWERING

Soil behaviour is controlled by the principle of effective stress and the use of drainage methods to control pore water pressure is of major importance in many practical situations. Successful building development on some types of poor ground will be strongly influenced by, and contingent on, appropriate drainage measures.

Principle

Lowering the ground water level will reduce pore pressures and increase the effective stresses within the soil. Thus a temporary lowering of the ground water level will over-consolidate the ground.

Methods and equipment

Some means of pumping water out of the ground is required. Methods of ground water lowering, including sump pumping, well-points, deep wells and horizontal drainage have been reviewed by Bell and Cashman (1985) and Preene *et al* (2000).

Testing

Monitoring of the piezometric level and settlement of the ground is an essential part of the treatment process.

Applications

The applicability of this improvement technique depends mainly on the permeability of the ground to be drained; Van Impe (1989) quotes a minimum value of $k = 10^{-8}$ m/s.

Performance

Tomlinson and Wilson (1973) have described the pre-loading of a 15 m deep end-tipped colliery spoil at Birtley in County Durham where, in addition to moving a 5.5 m high surcharge of colliery spoil across the site, the ground water level was lowered by 2.5 m. The option of using a 6 m ground water lowering was rejected due to the high cost of pumping and the possibility of detrimental effects on adjacent buildings. The site was to be used for a steel frame factory extension with a floor loading of 85 kPa.

Key features

1 The feasibility of the treatment method depends on the permeability of the soil.
2 The cost of pumping and the possibility of causing damaging settlement at adjacent buildings limit the applicability of the method.
3 The effluent pumped from the site may be polluted and so should be handled as contaminated waste.

8.5 VACUUM PRE-LOADING

In this form of pre-loading, the effective stresses are increased by reducing the pore water pressures rather than by increasing the applied stress.

Principle

Water is drained from the ground by the application of a vacuum. A hydraulic gradient towards the drains is created by reducing the pressure in the drains rather than by creating excess pore water pressure in the surrounding ground. For design purposes, vacuum loading generally is modelled as a pseudo-surcharge loading.

Methods and equipment

An impermeable membrane is placed over a granular filter layer and sealed into the clay layer at its edges. Vertical drains are usually installed in the clay layer. A vacuum pump is used to pump air out of the granular filter layer causing the clay to consolidate.

Testing

Monitoring of pore pressure and settlement of the ground should be undertaken.

Applications

The technique was developed in Sweden. Kjellman (1952) describing a large-scale field test using cardboard wicks as vertical drains at 0.5 m centres. The method is applicable for soft saturated clays where there is a high water table. The cost of electrical power needs to be considered.

Performance

At the Tianjin Port East Pier project in China, surcharge pre-loading of hydraulically placed silty clay caused fairly large movements, and minor slips occurred. Surcharging was temporarily suspended. Eventually the work was completed using vertical drains and vacuum pre-loading (Choa, 1994).

Vacuum pre-loading has been used on 480 000 m² of reclaimed land at Xingang Port in China (Shang *et al*, 1998) and on an area of 133 000 m² of soft clay for a waste water treatment plant at Kimhae in Korea (Park *et al*, 1997). A schematic arrangement for the method is shown in Figure 8.7.

Key features

1 Effective stresses are increased, with consequent consolidation of the soil, by reducing pore water pressures rather than by increasing the applied stress.
2 The method has an advantage on soft ground where a surcharge of fill could cause instability.
3 There are practical difficulties associated with maintaining the vacuum and reference should be made to case histories.
4 On contaminated sites, the discharge of effluent will present difficulties.

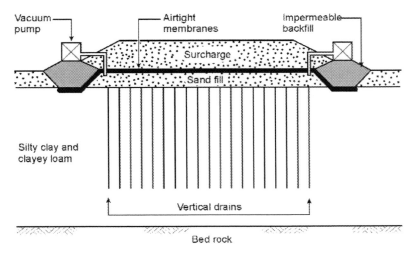

Figure 8.7   *Vacuum pre-loading (after Park* et al, *1997)*

8.6　ELECTRO-OSMOSIS

Hydraulic flow can result from an electrical gradient. When an anode and a cathode are placed in a fine grained soil and a voltage is applied, water flows towards the cathode at a rate that depends on the voltage. If water is removed at the cathode and not replaced at the anode, the moisture content is reduced and consolidation of the soil occurs.

Early development of this technique was carried out by Casagrande (1947). Electro-osmosis can be effective in dewatering silts and silty sands, but power requirements can be high and the method expensive. There have been many successful applications to soft foundations throughout the world, as well as a number of failures (American Society of Civil Engineers, 1978). However, the method has not been much used in the UK.

9 Improvement by stiffening columns

> Improvement is achieved principally by the installation of stiff vertical columns rather than by improving the properties of the ground itself. In some cases the process of installing the columns may also stiffen the surrounding soil. Where the columns are formed of granular material, their effectiveness is totally dependent on the support of the surrounding soil. Chemically stabilised soil columns have some inherent strength. Case histories of monitored field performance are used to illustrate what can be achieved by ground treatment in terms of improving soil properties, and limiting post-construction settlements.

9.1 APPLICABILITY OF STIFFENING COLUMNS

The objective of treatment by installing relatively stiff material in vertical columns is that the columns and surrounding soil form a composite system with a stiffness substantially greater than that of the untreated ground. There are different types of column that are used to stiffen the ground and these exhibit a wide range of behaviour.

- With granular columns, such as vibro stone columns, the strength of the column is totally derived from the lateral restraint provided by the surrounding soil.

- With lime columns and other types of stabilised soil columns, the columns have some inherent strength but the lateral restraint provided by the surrounding soil is also important.

- Vibro compacted concrete columns are analogous to piles with the loads being transmitted to a firmer stratum at depth.

Where stiffening columns are installed in a clay soil, the improvement is due to the stiffening provided by the column rather than any improvement to the clay itself and it is not so easy to characterise the improvement as it is, for example with the compaction of a sand. *In situ* tests will measure properties of the column or the surrounding soil, but may give little indication of the behaviour of the composite system. One approach is to relate the reduction in compressibility, expressed as the settlement ratio, s_r, (ie ratio of settlement with treatment to settlement without treatment), to the area ratio, A_r, defined as the proportion of the total area of the treated ground occupied by the columns.

9.2 VIBRO STONE COLUMNS

The ground treatment techniques most common in the United Kingdom are the deep vibratory processes collectively described as "vibro". The system originated in Germany to densify natural loose sands, as described in Section 7.2 on vibro-compaction.

Principle

The vibro system has been adapted for a wide range of soils and fills, including cohesive materials, by compacting gravel or crushed rock into the cylindrical void created by a vibrating poker. In clay soils the treatment will effect little or no improvement of the soil itself, but the stone columns will stiffen the ground.

Methods and equipment

A large vibrating poker, which normally has a diameter between 0.3 m and 0.45 m, penetrates to the design depth and the resulting cylindrical hole is filled with hard, inert stone, which is free of clay and silt fines. The stone is compacted in stages. The basic unit is usually about 4 m long and weighs approximately 2 tonnes. Extension tubes can be added to the basic unit to suit the depth of treatment and a total weight of 3 to 5 tonnes is common.

There are three principal systems of installing vibro stone columns: dry top-feed, wet top-feed and bottom feed systems. The majority of vibro stone columns in the United Kingdom are constructed by dry top-feed stone supply with the vibrating poker assembly suspended from a crawler mounted crane. The poker penetrates the ground under a combination of its own weight, the use of compressed air jets and vibration. After reaching the required depth, the vibrating poker is held in the ground for a short time and then withdrawn. A small charge of stone is tipped into the hole and the vibrating poker is lowered to compact the stone and interlock it with the surrounding soils. By adding and compacting successive small charges of stone, a dense column of stone is built up to ground level.

The wet top-feed system is used principally for the treatment of cohesionless soils below the water table or in weak silts and clays. The vibrating poker is suspended from a crane and penetrates the ground to the required depth under its own weight, with the aid of water jetting and vibrations. The water jetting flushes loose soil particles out of the bore, forming an annular space around the poker. Following formation of an open hole, the poker is kept in the ground and the water flow reduced to maintain circulation and stabilise the bore. Stone is introduced to the base of the hole via the annulus around the poker and compacted in short lifts. The wet process has problems associated with water supply, drainage ditches, settlement lagoons and final disposal of the effluent.

The introduction of bottom-feed guided pokers has extended the range of practical applications of vibro, including the use of a dry construction technique below the water table. A minimum undrained shear strength of 20 kPa is generally required when using the dry bottom-feed process. With this system, the vibrating poker has a stone supply tube attached down one side, which delivers stone to the tip of the poker. The poker penetrates the ground and the supply tube is charged with stone before continuing to the required depth under the combined action of the vibrations and self-weight, assisted by compressed air and an additional pull-down force. Additional stone is fed to the top of the poker by a travelling hopper system. The stone column is formed and compacted by lifting the vibrator, holding the lift for a short time to allow the stone to run, and then forcing the vibrator down on the charge of stone. This is repeated until a compact stone column is formed up to ground level.

The different vibro systems have been described by Greenwood and Kirsch (1983) and Moseley and Priebe (1993). A technical specification with notes for guidance has been published by Building Research Establishment (2000). Figure 9.1 shows stone columns being installed using the dry top-feed method.

The durability of stone columns could be affected either by the deterioration of the stone column material or by some reduction in the support to the column provided by the surrounding soil. The phrase "stone which is clean, hard and inert" conveys the spirit of the requirements for stone column material, but does not define acceptable criteria for the material type, grading, hardness and chemical stability which will largely determine the ability of the columns to fulfil the design requirements. The material used should be chemically inert and remain stable in the particular soil and ground water conditions at the site, making allowance for any changes, particularly in site hydrology, that might reasonably be anticipated. Limestone aggregate may not be suitable in acidic ground conditions, although this will depend on the pH level and the type of limestone available. Due regard should be given to the presence of any chemical contaminants identified in the site investigation. Detailed information on aggregates has been published by the Geological Society (Smith and Collis [eds], 1993).

Where weak soil deposits are deep, there may be practical or economic reasons for considering the use of partial depth treatment, but before adopting this approach it should be critically appraised. The minimum column length required to stiffen ground affected by the foundation loading needs to be evaluated and the likely behaviour of soil below the treated zone should also be considered. The introduction of stone columns could have adverse effects. For example, in loose unsaturated fills, stone columns could provide a pathway for the infiltration of surface water into the underlying untreated material and trigger collapse settlement.

In soft cohesive soils, design is often based on methods proposed by Hughes and Withers (1974) for the determination of the load-carrying capacity and the minimum column length required to prevent end bearing failure occurring at the toe of the column before bulging failure at a critical depth near the top of the column. The work of Baumann and Bauer (1974) can be used for settlement calculations, and design charts have been presented by Priebe (1995).

Figure 9.1   *Installation of vibro stone columns using the dry top-feed method (courtesy of Pennine Vibropiling Ltd)*

Testing

The stone column plate load test is the most widespread form of test carried out for quality control. Due to its size it can usually only influence the top part of the stone column. It can be used to control workmanship and consistency, at least in part, but cannot provide an assurance of performance requirements.

Performance testing can verify the degree to which ground improvement has been achieved and confirm that this meets the specified objectives. The behaviour of soil, stone column and foundation is interactive and complex. Testing a portion of the treated ground, which reflects the prototype in scale and geometry, is the most effective way of proving the performance of the treatment. On large sites where a considerable amount of treatment is proposed, sites with complex ground conditions, or marginal sites where the suitability of using stone columns requires verification, a trial of the treatment before the main treatment work may be appropriate. In these circumstances performance testing will play a vital role in developing the design of the vibro stone column treatment.

Performance test methods usually involve a large-scale load test carried out on a non-working foundation base using kentledge. Large-scale load test results are intended to indicate the actual foundation performance; such tests should be carried out wherever practicable and economically viable. Where ground conditions are judged to be particularly difficult, for example soft or weak natural soils or very variable fills, or where structures sensitive to settlement are to be supported, it is recommended that some form of performance testing be incorporated into the treatment specification. In cohesive soils a significant portion of total settlement will be related to consolidation and testing over a longer period may be necessary. As far as possible the test method should simulate the full-scale application, which may consist of rows or small groups of columns to support strip and individual pad foundations, columns on a grid pattern to support larger raft type foundations, or general widespread ground improvement to support earthworks. Large building and civil engineering applications will usually warrant performance testing. Larger sites may mean greater variability in ground conditions and will therefore require a number of tests to confirm uniformity of ground performance. Where widespread treatment is used to support extensive earthworks, a single large zone test involving the use of a trial embankment incorporating geotechnical monitoring systems may be helpful.

Large plate load tests are designed to span one or more stone columns and the intervening ground. This type of test may also be referred to as a *pad test* or *shallow maintained load test*, BS 1377: Part 9 (British Standards Institution, 1990) but should not be confused with a *zone test*, which refers to a test over a much wider and deeper zone of ground. The large plate load tests may comprise:

- a circular plate of diameter sufficient to load both the stone column and a soil annulus around the column
- a reusable rigid steel fabrication, which is bedded on sand or blinding concrete laid on the prepared surface
- a reinforced concrete pad cast-in-place on the prepared surface.

The test should be conducted at foundation formation level with the base of the excavation prepared by hand to minimise disturbance. The stiffness of a steel plate or concrete pad should be checked to be sufficient to ensure that a load applied centrally on the top surface is distributed across the underside. Loads may be applied by a jacking system, as in the column plate test, with reaction provided by heavy mobile plant for short-term tests. For longer tests direct loading can be applied by kentledge.

Bearing capacity and settlement performance are measured over a wider and deeper zone of ground in zone tests and facilitate comparison of the performance of the treated ground with design predictions. This may involve loading a full-scale structural foundation element such as a house raft foundation or a portion of ground with widespread treatment. In the case of widespread treatment load is often applied by temporary earthworks and may be part of a trial embankment study.

In situ penetration tests such as CPT, SPT and DP may be used where direct changes in properties of the soil, due to vibro stone column installation, can be measured and directly related to criteria set out in the contract documents or compared with pre-treatment test data. These techniques are principally of value in soils that are compacted by the vibratory action of the vibro poker, and may be used to confirm that the required improvement in engineering properties of the intervening ground has been achieved.

Applications

Over the last thirty years the system has been used widely on sites for low-rise buildings, particularly houses and light industrial units, and this is the main use in the United Kingdom. Typically, vibro is used to treat areas of shallow heterogeneous fill overlying less compressible soils. The dense stone columns reinforce the ground and should reduce total and differential settlements under foundation loading to acceptable values. The foundation design should recognise that the treatment results in stone columns at discrete locations with the intervening ground having a different stiffness. Possible effects of the introduction into the ground of vertical columns of high permeability should also be considered in the foundation design.

The durability of stone columns could be affected by the deterioration of the stone column material or by some reduction in the support to the column provided by the surrounding soil. The material should be chemically inert and remain stable in the soil and groundwater conditions at the site, making allowance for any changes, particularly in site hydrology, that might reasonably be anticipated. Where the soil surrounding stone columns contains organic material, biodegradation of the organic material can result in removal of support. Consequently, the method is not suitable for soils with a significant organic content.

Performance – widespread loading on fully penetrating columns

The simplest situation to examine is a widespread loading on a large number of fully penetrating columns. This has been investigated in various ways. Figure 9.2 shows the relationship between settlement ratio and area ratio derived from:

- an elastic analysis by Balaam and Booker (1981)
- an analysis by Priebe (1995) which assumed plastic yielding of the columns at constant volume
- some large-scale oedometer tests carried out by Charles and Watts (1983)
- centrifuge tests reported by Craig and Al-Khafaji.

Despite the large spread of the laboratory and analytical data plotted on Figure 9.2, in the range of practical interest for stone columns installed in clay soils with $0.25 < A_r < 0.35$, the data suggests that s_r values of the order of 0.5 are likely. This is confirmed by the limited field data described in Boxes 9.1 and 9.2

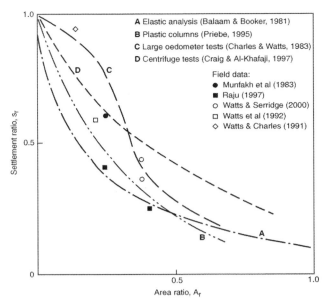

Figure 9.2 *Stiffening effect of stone columns*

Box 9.1 *Vibro stone columns for a wharf structure (after Munfakh et al, 1983)*

> The effectiveness of vibro stone columns installed in a deep deposit of soft clay was investigated through a full scale field trial in which a reinforced earth embankment modelled the proposed wharf structure in New Orleans. The average column diameter was 1.1 m and the columns were placed on a triangular pattern at 2.1 m centres, such that $A_r = 0.25$. It was estimated that $s_r = 0.6$.

Box 9.2 *Vibro stone columns at a road interchange (after Raju, 1997)*

> Vibro stone columns were installed in very soft cohesive soils at two highway interchanges in Malaysia. At Kebun there were marine clays and at Kinrara mine tailings consist of clayey silts with a fine sand content of 15 per cent. Settlements were measured on treated and untreated ground. At Kebun with $A_r = 0.23$ it was estimated that $s_r = 0.4$, and at Kinrara with $A_r = 0.4$ it was estimated that $s_r = 0.25$.

Performance – pad and strip loadings

The relationships shown in Figure 9.2 relate to stone columns that fully penetrate the full depth of a soft clay layer and are subjected to a widespread load. In situations where small numbers of stone columns are installed under individual footings or the columns are only partially penetrating, the relationship between s_r and A_r will be somewhat different. Priebe (1995) has developed charts that provide a means of making allowance for these factors.

Also, where stone columns are installed in soils such as heterogeneous fills, some improvement in the fill itself can be expected as well as the stiffening effect of the columns. The following field data relates to some of these different situations, particularly to stone columns under pad and strip footings.

The performance of a trial pad foundation on ground treated with partially penetrating vibro stone columns has been compared with the performance of a similar pad on untreated ground at the Bothkennar soft clay test site (Watts and Serridge, 2000). A 3 m × 0.75 m pad was founded at the base of the stiff natural clay crust on soft soil reinforced with two stone columns installed at 1.5 m centres corresponding to

$A_r = 0.38$. The columns were installed to 5.7 m below founding level, which was 1.2 m below original ground level. The 50 mm settlement measured for the treated pad was less than half the settlement that was recorded for an untreated pad over a similar period at the same applied bearing pressure. Over the applied stress range of 37 to 70 kPa, values of s_r from 0.35 to 0.43 were calculated.

Two 2 m square pad load tests were carried out on 4 m deep clay fill at Abingdon (Watts and Charles, 1991). Four stone columns were installed to the underlying gravel under one of the pads. Each pad was loaded with two sand-filled skips and settlement was measured over a six-month period. The properties of the clay fill samples were measured in the laboratory and the settlement of each test pad was estimated using values of m_v. Predictions from the laboratory tests suggested that the columns under the test pad had little effect in reducing total settlement. Further information about this site is given in Appendix A4.

Vibro stone columns have been used to treat between 3 m and 5 m of miscellaneous fill at a site in Bacup (Watts *et al*, 1992). Ash and stone fill overlies a mixed cohesive fill. Stone columns were installed through the fill to the underlying glacial till at 1.8 m centres, corresponding to $A_r = 0.21$. A 9 m × 0.75 m test foundation strip was subsequently loaded in three increments to a bearing pressure of 123 kPa. The performance of the strip was compared with a similar strip constructed and loaded on untreated fill. The effectiveness of the stone columns in reducing maximum settlement increased with applied load and there was a maximum value of s_r of 0.59. Further information about this site is given in Appendix A5.

Key features

1 The most commonly used ground treatment techniques in the United Kingdom are the various deep vibratory processes collectively described as "vibro". A wide range of fills and natural soils can be effectively treated.

2 The spacing of the columns and the area ratio, A_r, are key elements in the design of vibro stone columns.

3 In the range of practical interest for stone columns installed in clay soils with $0.25 < A_r < 0.35$, experimental data suggest that the settlement reduction ratio, s_r is likely to be of the order of 0.5, ie settlement is reduced to about one half of the settlement that would have occurred if the ground had not been treated.

4 Before adopting ground treatment based on partial depth treatment, the proposal should be critically appraised. In loose unsaturated fills, stone columns could provide a pathway for the infiltration of surface water into the underlying untreated soil and trigger collapse settlement.

9.3   DYNAMIC REPLACEMENT

Where the main result of dropping a heavy weight on to the ground surface is to punch pillars of granular fill into the ground rather than compacting the ground, it is preferable to describe the treatment as dynamic replacement rather than dynamic compaction.

Principle

Dropping a heavy weight on to the ground surface can be used to punch large pillars of imported granular fill into the ground. The difference between dynamic replacement and dynamic compaction is similar to the difference between vibro stone columns and vibro-compaction.

Methods and equipment

The equipment, comprising a tall rig and a drop weight, is identical to that used in dynamic compaction. In some soils the repeated impacts of the weight will cause substantial pore water pressures to develop and it will be necessary to allow sufficient time between the various phases of the treatment for these pore pressures to dissipate.

Testing

In applications on clay soils, the rate of dissipation of excess pore pressures set up by the treatment should be monitored.

Applications

The method has been used on saturated fine soils where it is unlikely that dropping a heavy weight will produce significant compaction of the ground.

Performance

There are not many examples of the use of dynamic replacement. Two case histories where some monitoring of performance was undertaken are given in Boxes 9.3 and 9.4.

Box 9.3 *Dynamic replacement of soft alluvial soil (after Charles and Watts, 1982)*

> When a soft alluvial soil at Guildford was treated by "dynamic consolidation", it was concluded that the most obvious result of heavy tamping was the formation of granular columns within the alluvial soil. Pore pressures were monitored and it was found that pulses of pore water pressure occurred during impact loading and generally there was a build-up of pore pressure during heavy tamping. Dissipation of pore pressures occurred fairly slowly over a two to three week period after both phases of tamping.

Box 9.4 *Dynamic replacement of dredged reclamation (after Hendy and Muir, 1997)*

> Dynamic replacement was used on a 40 m deep dredged reclamation for a petroleum storage facility in Hong Kong. Weights of up to 35 tonnes were dropped from heights of up to 40 m. Spacings of treatment points ranged from 6 m to 11 m. Performance under load was assessed from the results of a full scale proof load test and by hydro-testing the completed tanks.

Key features

1   With saturated fine soils it is unlikely that dropping a heavy weight will produce significant compaction of the ground.

2   Where dynamic replacement is used, it will be necessary to allow adequate time for excess pore pressures set up by tamping to dissipate between the successive phases of the treatment process.

9.4 STABILISED SOIL COLUMNS

In situ mixing of soils and stabilising materials such as cement, lime and bitumen has been used in road construction for many years. The mixing has been achieved by earthmoving machines and only very shallow depths of soil are affected. Deep stabilisation by the formation of stabilised soil columns is a more recent development:

- Deep foundation stabilisation of soft clay was first used in Sweden in 1967 and the method has been widely used in Scandinavia since 1975.
- From the early 1970s, deep soil mixing has been developed in Japan to improve the properties of cohesive soils to considerable depths; cement or lime have been used and depths of as much as 50 m have been treated; cement is now the primary agent (Terashi and Tanaka, 1981; Toth, 1993).

Principle

Certain soils can be improved by treating them with an admixture. The stabilising agent, which can be a liquid, slurry or powder, is physically blended and mixed with the soil. While some admixtures merely act as a binder, active stabilisers, such as lime and cement, produce a chemical reaction with the soil with consequent desirable changes in the engineering properties (Ingles and Metcalf, 1972). The properties of binders such as cement, quicklime, PFA and gypsum have been discussed by Esrig (1999). Cement and quicklime generate significant heat when they hydrate.

There has been confusion in the literature between lime columns and lime piles. The latter are not strictly a soil mixing technique. Glendinning and Rogers (1996) distinguish three types of deep lime stabilisation of which the first is a type of stabilised soil column:

1 Lime columns: deep vertical columns of intimately mixed lime and clay

2 Lime slurry: pressure injection; slurry is forced into the pores and fissures in the clay

3 Lime piles: cylindrical holes in the ground are filled with lime.

Methods and equipment

Dry mixing methods are commonly used in Sweden, Finland and Norway to form lime, lime/cement and cement columns (Broms, 1999). In forming lime columns, the mixing tool is first rotated down to a depth corresponding to the required length of the columns. The tool is then rotated and slowly withdrawn as unslaked lime is forced down into the soil by compressed air, applied through holes located immediately above the horizontal blades of the mixing tool. The unslaked lime is mixed *in situ* with soft clay using the special mixing tool (Broms, 1993). The diameter of the columns is 0.5 m to 0.6 m and columns of 15 m length have been installed. Lime/cement columns have the advantage that high strength can be obtained for soft clays, whereas lime columns have much higher permeability than lime/cement or cement columns.

In the deep soil mixing method developed in Japan, the two principal stabilising agents are lime and cement. The method has been used primarily to improve bearing capacity and settlement performance. It has been extended to control liquefaction under seismic loading. Stabilising agents such as cement slurry or quick lime are forced into the ground under pressure and mechanically mixed with soft clay soils down to depths of 50 m. The mixing equipment penetrates to the specified depth and is afterwards withdrawn with rotation. The layer to be improved is mixed twice while the hardening agent is added through a pipe during penetration and withdrawal, and is mechanically mixed *in situ* with the soft soil (Toth, 1993). The soil-cement column method has been applied in Japan to improve the ground under the foundations of detached houses (Kataoka *et al*, 1992).

A classification of deep mixing methods has been presented by Bruce *et al* (1999):

- binder – slurry (W) or dry (D)
- mixing mechanism – rotary (R) or jet assisted (J)
- location of mixing action – over length of shaft (S) or at end of mixing tool (E).

Lime cement columns used in Scandinavia correspond to D-R-E in this classification.

A major research project *EuroSoilStab*, with partners from Finland, Ireland, Italy, Netherlands, Sweden and the United Kingdom, is being funded by the European Commission (Ahnberg and Holm, 1999; Ilander *et al*, 1999). The principal aim is to develop design and construction methods for deep stabilisation of soft organic soils for road, rail and other infrastructure applications in Europe. Figure 9.3 shows the installation of stabilised soil columns in soft peaty soil.

Testing

In large projects, test columns are made before treatment begins, facilitating the design of the columns. The quality and strength of columns formed using the Swedish technique are generally investigated using column penetrometer tests (Halkola, 1999). A mechanical penetrometer equipped with two vanes is pushed into the centre of the column without rotation, and compressive strength is measured at the upper end of the penetrometer rod. The cross-sectional area of the column penetrometer is 10 times as large as that of a standard CPT cone and its use is usually restricted to columns with $c_u < 200$ kPa. Because of the small area of the cone, the CPT requires a much larger number of columns to be tested in order to obtain a reliable estimate of strength.

Figure 9.3   *Installing stabilised soil columns (courtesy of Building Research Establishment Ltd and Keller Ground Engineering Ltd)*

Applications

Lime columns have been used to improve the bearing capacity of soft clay and to reduce settlement. They have been used widely in Scandinavia in the following applications (Broms, 1993):

- to improve the total and differential settlement of light structures (one- to two-storey buildings) (Bredenberg and Broms, 1983)

- to increase the settlement rate and control the settlement of relatively heavy structures

- to improve the stability of embankments, slopes, trenches and deep cuts

- to reduce the vibrations from, for example, traffic, blasting, pile driving.

Glendinning and Rogers (1996) have reported the use of lime piles to reinforce soft clay soils in Russia, Japan and China. Lime piles have been used in the United States of America and elsewhere to stabilise slopes and trial applications for this purpose have been carried out in the United Kingdom. The stabilisation mechanism concerns the migration of calcium ions from the pile into the surrounding clay and its subsequent stabilisation by lime-clay reaction, but dehydration of the clay is also often an attributed effect. The use of this type of ground treatment for improving the stability of ageing motorway earthworks has been reviewed by West and Carder (1997).

Deep soil mixing can be used in the remediation of contaminated land and can involve both stabilisation and solidification processes. The objective of stabilisation is to reduce the leachability and mobility of the constituents. Solidification aims to reduce fluidity and friability and prevent access by mobilising agents.

Performance

Treatment by soil mixing, which involves irreversible chemical reactions, is generally regarded as a form of permanent ground improvement. For lime columns, Ahnberg *et al* (1989) have discussed the effect of different curing conditions on the increase in strength with time after installation. Nevertheless, chemically modified soils could be subject to deterioration. For example, changes in site hydrology can have a detrimental effect on such soils and the chemical effects could change with time.

With stiff stabilised soil columns a given reduction in compressibility of the composite system should be achieved with a smaller area ratio than that required with stone columns. The field measurements presented in Boxes 9.5 and 9.6 give some confirmation of this, but the case described in Box 9.7 is more comparable with field data from stone column installations.

Box 9.5   *Lime columns in soft clay (after Bredenberg, 1983)*

Lime columns 0.5 m in diameter and at 1.3 m centres, corresponding to A_r = 0.13, were used to stiffen a soft clay at a cargo terminal in Stockholm, Sweden. Monitoring of treated and untreated ground gave s_r = 0.5 after nine months, but it was expected ultimately to be significantly smaller than this as the presence of the columns should have speeded up the rate of consolidation as well as reducing its magnitude.

Box 9.6   *Lime columns in hydraulic fill (after Soyez et al, 1983)*

Loading tests on lime columns installed in hydraulic fill, in Paris, France, show similar results to those described in Box 9.5. With columns at 1.5 m centres, corresponding to A_r = 0.09, s_r = 0.5. However with columns at 2.0 m centres, corresponding to A_r = 0.05, there was little reduction in settlement.

Box 9.7   *Stabilised soil columns in silty clay with peat layers (after Koehorst and van den Berg, 1999)*

At the Hoekse Waard test site in the Netherlands, 0.6 m diameter stabilised soil columns were installed at 1.1 m centres on a triangular grid (A_r = 0.27) through a silty clay with peat layers. The columns were formed using the dry method and the applied binder was a mix of 80 per cent blast furnace cement and 20 per cent anhydrite. A 5 m high embankment induced 1.1 m of settlement in the treated ground compared with 2 m of settlement in ground without the columns (s_r = 0.55).

Key features

1   Some soils can be improved by treating them with an admixture, which is physically blended and mixed with the soil. Active stabilisers, such as lime and cement, produce a chemical reaction with the soil with consequent desirable changes in the engineering properties. This is a rapidly developing area of ground treatment.

2   With stiff stabilised soil columns a particular reduction in compressibility of the composite system should be achieved with a smaller area ratio than that required with stone columns.

9.5   VIBRO CONCRETE COLUMNS

This technique was developed in Germany and introduced into the United Kingdom in 1991. Although it might be considered to be a piling method rather than a form of ground treatment, it has been included in the report because of its similarity to vibro stone columns (Section 9.2). It is probable that use of the technique will expand.

Principle

The concrete columns are designed to transmit structural loading to a suitable underlying bearing stratum and thus are effectively acting as piles.

Method and equipment

Vibro concrete columns (VCCs) are installed using a modified guided bottom-feed vibro rig and are formed by pumping concrete into the void formed by the poker penetrating very soft or organic soils. The installation of VCCs is shown in Figure 9.4.

Testing

The testing of vibrated concrete columns is carried out in a similar way to pile testing. Both static and dynamic tests are employed.

In the static test a high-strength reinforced concrete loading cap is cast onto the head of the column and load is applied by jacking against a reaction frame loaded with heavy kentledge. Where columns form the support for widespread loading, often incorporating a granular load transfer platform reinforced with geogrids (see Appendix A6), a zone test may be carried out in which several columns and the intervening ground are loaded to simulate the working situation.

Dynamic testing of concrete columns is increasingly used, often to supplement the more expensive static tests, which can be used as reference tests on larger sites. Dynamic testing requires a strong head cast onto the concrete column as in static testing and the column is impacted with a purpose made drop weight. Special instrumentation measures the dynamic response of the column, which is correlated with static test performance. Guidance on integrity testing in piling practice has been presented by Turner (1997).

Applications

Vibro concrete columns are being used in soft soil ground treatment. The technique is being used not only to treat very soft and organic soils, which are unsuitable for stone column application, but also some denser soils.

Performance

The method has more in common with piling than with ground treatment; this is confirmed by monitored field performance. In applications involving embankments a load transfer platform is often required. An example of this is described in Box 9.8.

Key features

1   Vibro concrete columns are installed using methods similar to those used in the installation of vibro stone columns.

2   Vibro concrete columns are designed to transmit structural loading to a suitable underlying bearing stratum and thus are effectively piles.

Box 9.8   *Vibro concrete columns with load transfer platform (after Maddison et al, 1996)*

The toll plaza for the Second Severn Crossing involved the construction of an embankment up to 6 m high over highly compressible peat and clay soils. The foundation system comprised vibro concrete columns (VCCs) at 2.7 m maximum spacing on a triangular grid founded in sand and gravel deposits, and a load transfer platform of granular fill incorporating low strength geogrids. At mid span settlements were typically 20 mm to 40 mm. The maximum observed settlement of the load transfer platform of 75 mm occurred at the west of the site where the VCCs were founded in silty sand soils; this was attributed to a lower bearing capacity of the soils in that location. Without ground treatment, settlements of 1 m had been anticipated; the VCCs were not so much stiffening the compressible peat and clay soils, but rather acting as piles transmitting the loads to a firmer underlying stratum. Further details of this case history can be found in Appendix A.6.

Figure 9.4   *Installing vibro concrete columns (courtesy of Building Research Establishment Ltd)*

10 Current capabilities and recommendations for good practice

The proportion of sites where ground treatment is required prior to construction can be expected to increase and there is, therefore, a need for reliable methods for assessing the effectiveness of ground treatment processes. In this final section of the report, conclusions are drawn on current practice in assessing ground properties and long-term performance. Recommendations are given for good practice and for future research.

10.1 CURRENT PRACTICE

Most ground treatment in the United Kingdom is on filled ground. The fills that require treatment may be granular, clayey or organic, and include opencast backfill, colliery spoil, demolition materials and old domestic refuse. A significant, if much smaller, number of treatments are carried out on natural soft clay and organic soils. Current emphasis on sustainability and the reuse of brownfield sites make it very probable that the proportion of land requiring ground treatment prior to development will increase.

Many of the ground treatment techniques, which are currently available, involve densification by surface loading (pre-loading with a surcharge of fill, dynamic compaction, rapid impact compaction) or by subsurface methods (vibro-compaction). Other techniques involve physical reinforcement or chemical modification (vibro stone columns, vibro concrete columns, lime columns and lime piles, deep mixing). Vibro stone columns are the most commonly employed method of ground treatment in the United Kingdom.

Various methods are used to measure or estimate soil properties including *in situ* tests such as SPT, CPT, DP, field vane, field loading tests and geophysical testing. Nevertheless, in many cases it is difficult to predict long-term performance on the basis of the test results.

While most ground treatment appears to be successful in the sense that there are few reported cases of failure, in many cases little may have been required of the treatment process. The success rate may therefore be misleading when a major improvement in soil properties is essential to the success of the development.

The research project, on which this report is based, identified some shortcomings and areas of weakness in assessing ground properties and the long-term performance of treated ground. The application of ground treatment has often suffered from an inadequate understanding of ground treatment processes and lack of appreciation of what properties are required from treated ground by clients and those responsible for procuring and specifying ground treatment.

This has resulted in and been exacerbated by factors including:

- insufficient or inappropriate site investigation
- incorrect diagnosis of problem
- incorrect choice or inappropriate procurement of treatment
- inappropriate or unrealistic specifications

- poor execution of treatment
- insufficient or inappropriate testing
- inadequate liaison between those responsible for, respectively, ground treatment and structural design.

10.2 PREDICTION OF PERFORMANCE

During treatment and immediately following completion of treatment, testing should be designed to check and, it is hoped, to confirm that a specified improvement in soil properties has been achieved. Use of the observational method, see CIRIA Report 185 (Nicholson et al, 1999), enables changes to be made during the treatment process in order to achieve the required improvement. However, a failure to achieve acceptable performance of the treated ground due to excessive post-construction settlement, may only become apparent in the medium to long-term, which could be very difficult to identify in the short-term.

A thorough understanding of the treatment process and its effect on the behaviour of the ground is needed. When examining performance requirements and methods of predicting performance, a realistic outlook should be based on past experience and the monitored performance of similar types of treated ground. Without such a basis, there is a likelihood that unrealistic requirements and over-optimistic predictions of performance will be made. In many cases a principal objective of ground treatment will be to increase the stiffness and strength of the ground. In other situations the elimination of collapse compression on inundation will be a key requirement. At an early stage in the design process it is necessary to know the order of magnitude of the improvements that can be achieved by different treatment methods.

Treatment methods that provide improvement through the full depth of a fill deposit are likely to exhibit a more reliable long-term performance than those methods in which only the upper layers are treated. Where only partial depth treatment is practical or economically viable, there is the possibility that, although the treatment is durable, the performance is unsatisfactory because of deficiencies in the ground below the treated depth. Usually in this type of situation, the greater the depth and stiffness of the treated zone, the lower will be the risk of long-term problems associated with differential settlement. Where a treatment method consists of some form of loading applied to the ground surface, it is important to know the depth of effectiveness of the treatment. Field performance data provide some guidance for both dynamic compaction and pre-loading with a surcharge of fill.

An assessment of long-term performance at the design stage has to be based on the measured properties of the untreated ground and an evaluation of the modification of those properties, which will result from ground treatment. Performance should be estimated for realistic loading conditions and the drainage regime should be adequately characterised. Some confirmation of the prediction of long-term behaviour can come from soil properties measured during or immediately following ground treatment.

One possible cause of unsatisfactory performance in the long-term relates to deterioration processes within the treated ground. The nature of some processes means that they will be inherently less stable than others. Long-term ground movements could still occur even though the treatment process is durable and for settlement-sensitive buildings this could be the crucial performance indicator. Durability is conveniently examined under the three main forms of remediation, which have been identified namely, densification, reinforcement and mixing:

- Soils are not elastic materials and densification produced by compaction and consolidation processes is unlikely to be reversed: most soils that have been densified will remain in that dense state.

- The durability of stone columns could be affected either by the deterioration of the stone column material or by some reduction in the support to the column provided by the surrounding soil.

- Treatment by soil mixing, which involves irreversible chemical reactions, is generally regarded as a form of permanent ground improvement; nevertheless, chemically modified soils could be subject to deterioration.

10.3　RECOMMENDATIONS FOR GOOD PRACTICE

This CIRIA report has drawn together lessons from case studies and the experience of practitioners. It should be remembered that, for construction purposes, poor ground can be improved but bad ground cannot be made into good ground. The following recommendations relate to aspects identified to be where improved practice is needed, and for which the publication of this report should promote industry awareness.

1　*Improved diagnosis of deficiencies in load-carrying properties of the ground.* The diagnosis of such deficiencies is the vital first step in selecting an appropriate technique for ground improvement. Routine foundation design is often based on providing adequate bearing capacity and making sure that the bearing pressures do not produce unacceptable settlement. However, much of the ground treatment work in the United Kingdom is currently on filled sites and with fill materials the cause of unsatisfactory long-term performance is more likely to be associated with other factors, such as long-term creep, biodegradation of organic matter or collapse settlement through an increase in soil moisture.

2　*Improved understanding of the effects of ground treatment and what can be achieved.* The type of ground improvement best suited to alleviate, or preferably eliminate, the deficiencies in load-carrying properties needs to be identified. The nature and degree of improvement that can be achieved should be evaluated. For example, if depth of improvement is important, surface compaction should be designed to achieve the improvement over the required depth. If improvement by stiffening columns is the aim, as in the use of vibro stone columns, the columns should be spaced appropriately to support the applied structural loads.

3　*Closer links between ground treatment and foundation design.* The ground treatment design should be based on a realistic assessment of loading. Close liaison between the structural and ground treatment designers is needed to avoid misunderstandings concerning loadings and unrealistic expectations of the treatment. Clear allocation of responsibilities is also required.

4　*Development of appropriate quality management procedures for ground treatment and its assessment.* Comprehensive and updated standard specifications, based on best practice, are needed to ensure high technical standards are achieved in a climate of fair competition. This should produce the best technical outcome and value for money for the client in terms of whole-life costs.

5　*Use of testing techniques relevant to the principal causes for concern for long-term performance of the treated ground.* The load-bearing capability and settlement characteristics of the treated ground should be confirmed with tests that model as closely as practicable the scale and magnitude of the actual foundation loading. Where long-term performance is likely to be affected by other factors, for example, in deep fill where collapse settlement on inundation may be the major hazard, an assessment of increase in density with depth resulting from the treatment will be relevant.

6   *Risk management.* Not all risk for subsurface conditions can be avoided or eliminated and, where ground conditions are poor, the risks for those involved in the development can be high. Ground treatment should be properly designed and executed, based on an adequate understanding of the deficiencies of the ground, as well as the required ground behaviour. Ground treatment should have a major role in the promotion of developments where the risks are identified, managed and controlled. The previously listed five items should all assist in risk management, but there is clearly scope for a more explicit application of risk analysis and risk assessment procedures.

10.4   RECOMMENDATIONS FOR RESEARCH

The background information for assessing the effectiveness of ground treatment is provided by well documented case histories of long-term performance. However, this type of information is not readily available and current developments in methods of procurement mean that it is less likely to be available in the future. It is necessary to make full use of all the data that are available.

There is a need for research based on case history observations and reporting. More analysis of past performance data, including structural performance, is required. It would be helpful to analyse available data to examine the incidence of ground treatment that has not given satisfactory long-term behaviour. While this may seem inappropriate in a climate of litigation and a culture of blame, negative as well as positive feedback into the design, application and testing processes is essential if improvements are to be achieved. Development of systematic risk assessment models would be useful.

Full-scale demonstration projects incorporating long-term post-construction monitoring can yield valuable performance data. The empirical nature of ground engineering design, particularly in respect of extremely variable and poorly categorised ground types often found on brownfield sites, means that this is the most likely type of investigation to result in meaningful data, with supporting information from laboratory testing and numerical modelling.

Appendices – case histories

Throughout the report reference has been made to case histories because these provide valuable, and sometimes the only, information on which the applicability, performance and durability of ground treatment can be judged. The measurement of engineering properties in relation to long-term performance of treated ground can usually only be assessed through the study of specific case histories. Table A.1 lists the case histories and their location within the report. Several of the more important case histories are described in greater detail in the appendices.

A1   Refuse fill – dynamic compaction

A2   Clay fill – pre-loading

A3   Low permeability natural soil – pre-loading

A4   Mixed clay fill – vibro stone columns

A5   Miscellaneous fill – vibro stone columns

A6   Highly compressible soils – vibro concrete columns

Table A.1   *Summary of case histories*

Box	Section	Location of site	Treatment method	Ground conditions	Development
5.1	5.2.3 (5.3.1)	Peterborough, UK	No treatment	Lagoon PFA	Township
	5.3.1 (5.2.3)	Peterborough, UK	No treatment	Lagoon PFA	Township
	5.3.1 (7.2)	Wythenshawe, UK	Vibro-compaction	Sand	Housing
6.1	6.1	Bristol, UK	Vibro stone columns Vibro concrete columns	Soft clay	Road embankment
6.2	6.6	Warrington, UK	Vibro stone columns	Alluvial soils	Factory
6.3	6.6	West Midlands, UK	Vibro stone columns	Opencast backfill	Housing
6.4	6.6	Peterborough, UK	Vibro stone columns	Fill	Service station
	7.2	Belawan, Sumatra	Vibro-compaction	Sand fill	Port development
	7.2	Chek Lap Kok, Hong Kong	Vibro-compaction	Sand fill	Airport
7.1	7.2 (5.3.1)	Wythenshawe, UK	Vibro-compaction	Sand	Housing
7.2	7.2	Flintshire, UK	Vibro-compaction	Sand	Bridge abutment
	7.3	Changi, Singapore	Dynamic compaction	Sand fill	Airport
	7.3	Pulau Ayer Merbau, Singapore	Dynamic compaction	Sand fill	Warehouse
	7.3	Ashuganj, Bangladesh	Dynamic compaction	Sand fill	Factory
7.3, 7.4	7.3	Snatchill, Corby, UK	Dynamic compaction	Clay fill	Housing
7.5	7.3	Redditch, UK	Dynamic compaction	Old refuse	Road
7.6	7.5	Peterborough, UK	Compaction in thin layers	Clay fill	Housing
	7.5	Not known	Compaction in thin layers	Clay	Distribution centre
	8.2 (A.2)	Snatchill, Corby, UK	Pre-loading	Clay fill	Housing
	8.2	Staffordshire, UK	Pre-loading	Clay and shale fill	Housing
8.1	8.2	Irlam, UK	Pre-loading	Soft clay	Railway embankment
	8.3	Orebro, Sweden	Pre-loading with drains	Soft clay	Test site
8.2	8.3	Antoniny, Poland	Pre-loading with drains	Peat and gyttja	Embankment
	8.3	Monnickendam, Netherlands	Pre-loading with drains	Clay and peat	Embankment
	8.4	Birtley, County Durham, UK	Ground water table lowering	Colliery spoil	Factory extension
	8.5	Tianjin Port, China	Vacuum pre-loading	Silty clay fill	Pier
	8.5	Xingang Port, China	Vacuum pre-loading		Port development
	8.5	Kimhae, Korea	Vacuum pre-loading	Soft clay	Waste Water Treatment Plant
9.1	9.2	New Orleans, USA	Vibro stone columns	Soft clay	Wharf
9.2	9.2	Kebun, Malaysia	Vibro stone columns	Soft clay	Highway interchange
9.2	9.2	Kinrara, Malaysia	Vibro stone columns	Clayey silt fill	Highway interchange

Table A.1 *Summary of case histories (continued)*

Box	Section	Location of site	Treatment method	Ground conditions	Development
	9.2	Bothkennar, UK	Vibro stone columns	Soft clay	Test site
	9.2 (A.4)	Abingdon, UK	Vibro stone columns	Clay fill	Two storey steel frame building
	9.2	Bacup, UK	Vibro stone columns	Miscellaneous fill	Test site
9.3	9.3	Guildford, UK	Dynamic replacement	Soft clay	Road
9.4	9.3	Hong Kong	Dynamic replacement	Fill	Petroleum storage facility
9.5	9.4	Stockholm, Sweden	Lime columns	Soft clay	Cargo terminal
9.6	9.4	Paris, France	Lime columns	Hydraulic fill	Test site
9.7	9.4	Hoekse Waard, Netherlands	Stabilised soil columns	Soft clay with peat layers	Embankment
9.8	9.5	Second Severn Crossing, UK	Vibro concrete columns	Peat and clay	Toll Plaza
	A.1	Cwmbran, UK	Dynamic compaction	Refuse fill	Warehouse
	A.2 (8.2)	Snatchill, Corby, UK	Pre-loading	Clay fill	Housing
	A.3	Port Talbot, UK	Pre-loading	Soft clay	Road
	A.4 (9.2)	Abingdon, UK	Vibro stone columns	Clay fill	Two storey steel frame building
	A.5	Bacup, UK	Vibro stone columns	Fill	Load test
	A.6	Second Severn Crossing, UK	Vibro concrete columns	Soft clay	Toll Plaza

A.1   REFUSE FILL – DYNAMIC COMPACTION

Reference: Downie and Treharne (1979)

Private communication: Ian Statham, Ove Arup and Partners (2000)

Commentary

This typical reclamation project involved the treatment of miscellaneous domestic and industrial waste deposits of variable depth to allow long-term serviceability of warehouses over a 20-year design life. *In situ* testing was effectively used for control of the treatment and to compare pre- and post-treatment ground conditions. Monitoring of structural settlement over a period of 10 years has given a valuable indication of long-term behaviour of the treated ground.

- Settlement of one unit has been relatively uniform but was still continuing at a significant and constant rate after 10 years. It is likely that this is principally due to biodegradation of waste. While densification should improve most engineering properties, it is unlikely to eliminate biodegradation and the associated volume reduction.

- Distortion of some units has occurred. In one case this was caused by substantial overloading of part of a floor area. This is an inherent problem where the end use of a structure may change during its working life and the designer has no control over such changes. To design for all possible contingencies would lead to over-design, unacceptable construction costs and may not be technically feasible when treating poor ground.

Required ground behaviour (Section 2)

Reclamation of a site at Cwmbran was carried out for a warehouse development. The development comprised seven warehouse units, each 40 m × 48 m in plan and 6 m high with a roof supported by perimeter and central columns. The warehouses were designed for floor loadings of 15 kPa and to have a design life of 20 years. Site layout took account of quarry geometry to avoid extreme variations of fill depth beneath the individual units. The principal problems to be addressed were excessive total and differential settlements of the fill and, thus, long-term serviceability of the structures.

Deficiencies in ground behaviour (Section 3)

Between 1962 and 1971 a quarry from which brick-making materials had been removed, was filled with a mixture of domestic and industrial refuse. Site investigation showed the quarry generally to be 5 m to 10 m deep, with steep sides and three deeper trenches reaching to an estimated maximum depth of 16 m some 2 m to 3 m below the water table. The industrial refuse was reported to include a substantial amount of iron and steel slag, including gravel size fragments and some larger lumps, reputedly up to 10 m³.

Engineering properties (Section 4)

Trial pits discovered numerous fragments of slag and confirmed that the tip had been sealed by a clay blanket, generally 0.5 m thick. They also showed that the refuse was 40 per cent domestic and 60 per cent industrial waste; the domestic refuse being mainly ashes, glass bottles, plastic containers, metal objects, leather, wood, cloth and some clay and sand in a black organic matrix. It was estimated that the plastic content amounted to less than 10 per cent of the total volume and that the readily degradable organic material was less than 10 per cent. The industrial refuse was mainly sand, masonry and concrete with some metal and wooden objects, also traces of rubber tyres and plastic sheets. There was little sign of decay.

Measurement of engineering properties (Section 5)

Menard pressuremeter tests and standard penetration tests were performed throughout treatment to measure the *in situ* parameters of the refuse. Figure A.1.1 shows typical pre- and post-treatment values of the pressure limit and modulus of deformation from the pressuremeter tests performed at Unit 1 and also typical standard penetration tests values measured before and after treatment at blocks 5, 6 and 7. Both methods confirmed significant improvement as treatment progressed. The pressuremeter results were more consistent and showed a greater improvement ratio; approximately 200 per cent. Levels taken after each phase of treatment indicated total enforced settlements ranging from 0.35 m to 1.0 m and this corresponded to a volume decrease in the refuse of approximately 10 per cent. Three pad load tests were carried out using kentledge to check the ground performance against the specification. The results showed settlements of 12.6 mm, 11.1 mm and 13.9 mm under the specified loading of 150 kPa in close agreement and within the required limit of 20 mm.

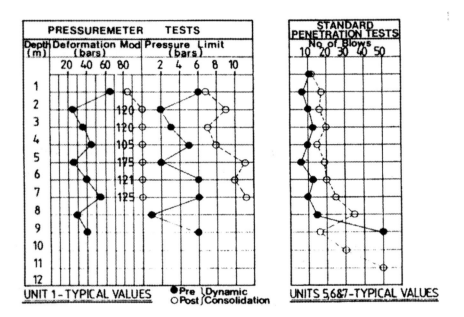

Figure A.1.1 *Control testing results at Cwmbran (after Downie and Treharne, 1979)*

Provision of ground treatment (Section 6 and 7.3)

Dynamic compaction used a 15 tonne weight dropped from 17 m on a 6 m square grid. The treatment was phased into a number of passes with control testing between each. Up to 6 passes proved necessary to achieve the specified compaction. The total input energy varied from 120 to 250 tonne m/m². Soft areas were encountered where there were concentrations of household refuse in plastic bags and large craters were backfilled and compacted using extra energy. Where the causes of the weakness were close to the surface, refuse was excavated, dispersed, and the excavation filled with granular material. The variability of the ground surface meant a granular working blanket was required. The blanket of steelwork slag had a dual function; it provided a stable working platform and it helped to reinforce the upper few metres of refuse when it was driven in under repeated impacts. By the completion of works almost 0.5 m thickness had been added over the entire working surface of the site. Occupation of unit 1 commenced in November 1976 with steel storage. Unit 2 was occupied in the summer of 1976 and is used for lightweight storage. Lightweight storage began in unit 3 in late 1977.

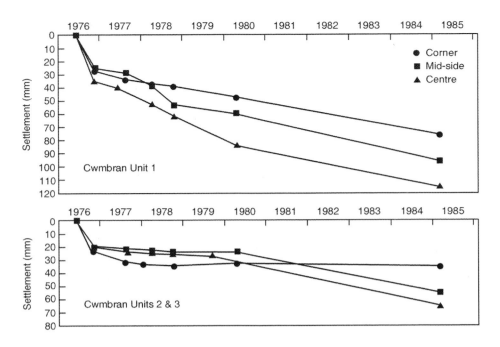

Figure A.1.2 *Long-term settlement of units at Cwmbran*

Figure A.1.2 shows a summary of the settlements of unit 1 and units 2/3 during the period from completion in May 1976 to April 1985. Long-term settlement of unit 1 is relatively uniform but was still continuing at a constant rate of about 7 mm per year at the end of 10 years. Some distortion of units 2 and 3 has occurred. When inspected in 1992, the buildings were still serviceable although some minor damage to one unit had resulted from substantial overloading of part of the floor area.

CLAY FILL – PRE-LOADING

References: Charles *et al* (1978)
                   Burford and Charles (1991)

Commentary

A range of ground treatment techniques, including an attempt to use inundation to precipitate collapse in untreated fill prior to building development, were investigated in this unique case study. The long-term performance of the treated ground and the houses built on it has been measured over a period of 25 years and compared directly with an area of untreated ground.

- Pre-loading proved the most effective treatment in terms of both depth of effectiveness and reduction of long-term settlement. The pre-loading also produced the most uniform improvement as indicated by long-term differential movements of houses built on the treated ground.

- The largest movements, in terms of both total ground settlement and differential settlement of houses, were measured in the area where inundation was carried out from surface trenches. The trial highlighted the unpredictability of the collapse effect, even over a relatively small area, and the measurements showed that the effect, once triggered, could carry on in an uncontrolled manner for many years. The technique was unsatisfactory as a method of treatment but the experiment clearly demonstrated that water penetrating the backfill from the ground surface could be a major hazard to construction on the site.

Required ground behaviour (Section 2)

The expansion of Corby new town involved housing and industrial development on land that had previously been worked for ironstone by opencast mining methods. Restoration of the Snatchill site was completed in 1970. The principal problem to be addressed was differential settlement.

Deficiencies in ground behaviour (Section 3)

Boulder clay and oolitic limestone overburden had been stripped by a walking dragline excavator and dumped so that the upper part of the 24 m deep unsaturated backfill was predominantly stiff clay. The lower part of the backfill consisted mainly of oolitic limestone. The water table has remained below the level of the backfill.

The principal deficiencies of the fill were long-term creep settlement due to self-weight and the potential for collapse settlement due to wetting.

Engineering properties (Section 4)

The properties of the upper part of the backfill were of principal interest. As expected, the stiff clay fill was very variable.

The clay fill had an undrained shear strength of about 100 kPa. Moisture contents varied between 7 per cent and 28 per cent with a mean value of 18 per cent. The corresponding range of densities was 1.5 to 1.8 Mg/m^3, with a mean value 1.7 Mg/m^3. A particle size analysis indicated that 54 per cent was finer than 0.075 mm, and the clay fraction was 19 per cent. Typical values for plastic and liquid limits were 17 per cent and 28 per cent respectively.

Measurement of engineering properties (Section 5)

Boreholes were drilled through the full depth of the fill and samples recovered for laboratory testing. Magnet extensometers were installed in these boreholes to enable settlement to be measured at different depths within the fill. The behaviour of the fill was monitored prior to treatment.

Provision of ground treatment (Section 6 and 8.2)

An area was preloaded with a 9 m high surcharge of fill. The surcharge was placed by towed scrapers over a three-week period in 1975. This is shown in Figure A.2.1. It was left in position for a month and then removed. Similar size areas were treated by dynamic compaction and inundation of the fill from 1 m deep water-filled trenches. A fourth area remained untreated.

Figure 8.2 shows that most of the settlement occurred as the surcharge was being placed, the mechanism being the closure of macro voids between the stiff clay lumps, rather than consolidation of the clay lumps themselves. The movements measured while the surcharge was left in position were much smaller. A small amount of heave occurred as the surcharge was removed. The stresses produced by pre-loading were much greater than those subsequently applied by house foundation loads.

Figure A.2.1 *Placing a 9 m high surcharge (courtesy of Building Research Establishment Ltd.)*

Settlements at different depths measured by four magnet extensometers located within the pre-loaded backfill are plotted in Figure A.2.2. The settlement versus depth profile suggests that the surcharge was effective down to a depth of 10 m. Similar measurements showed effective treatment depth for dynamic compaction was about 5 m. The effect of inundation was variable and unpredictable in terms of magnitude, space and time. The average settlement induced by each treatment method is recorded in Table A.2.1.

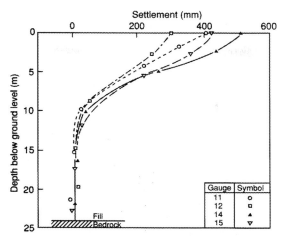

Figure A.2.2 *Settlement induced at depth by pre-loading at Corby (after Charles et al, 1978)*

Table A.2.1 *Average settlements (mm) induced by ground treatment*

	Dynamic compaction	Inundation	9 m high surcharge
At surface	240	100	410
At 4 m depth	90	40	230
At 10 m depth	<10	<10	40

One detached, one terraced and one semi-detached building were constructed on each of the treated areas and two semi-detached and one detached building on the untreated area. The houses were built to the contractor's standard design with cavity walls on trench fill foundations 375 mm × 900 mm deep with the top of the concrete 75 mm below ground level.

House building commenced in August 1975 and was completed by May 1976. Levelling points were grouted into the course of brickwork immediately above damp-proof course (DPC) level.

Movements measured at DPC level during, and subsequent to, construction show settlements have been smallest in the preloaded area and the performance of the houses built on this backfill has been very satisfactory (Figure 8.3).

The settlement of the houses built in the untreated backfill has been surprisingly small. However, houses built on untreated backfill in another part of Corby settled up to 180 mm in 10 years.

Total and differential settlements of houses on fill pre-inundated from surface trenches at the Snatchill site were substantially larger than the preloaded area or the area treated by dynamic compaction and clearly indicate that water penetrating the backfill from surface trenches could be a major hazard to construction on the site.

Maximum deflection ratios for the houses on the different areas are given in Table A.2.2 (sagging is positive, hogging is negative).

Table A.2.2 *Deflection ratios for houses on different areas*

Treatment area	Δ mm	L m	Δ/L
Inundation	-11	19.2	-0.57×10^{-3}
Surcharge	-3	19.2	-0.16×10^{-3}
Untreated	2	12.8	0.16×10^{-3}
Dynamic compaction	-3	19.2	-0.10×10^{-3}

LOW PERMEABILITY NATURAL SOIL – PRE-LOADING

Reference: Jones *et al* (1995)

Commentary

The use of extensive field trials has provided data to feed back into a design model for treatment by surcharging. The aim was not to eliminate settlements, but rather to restrict movements to within acceptable limits and accurately predict the extent and duration of post-construction movements. The trial helped to refine the design details for the surcharge embankment. The use of field instrumentation clarified the ground behaviour under loading and unloading conditions.

- Medium-term monitoring (five years) has enabled long-term behaviour (25 years) to be predicted with greater confidence and the results indicate a substantial decrease in predicted movements. This has been confirmed by monitoring ground where pre-loading treatment could not be carried out. The approach has enabled the development programme to be met.

- The case illustrates the use of the simple concept of pre-loading ground to increase stiffness. It highlights the need to understand the geotechnical processes and use them to achieve the desired outcome for a particular project.

Required ground behaviour (Section 2)

A major development near Port Talbot was planned on a site originally at about 4 m above Ordnance Datum (AOD). In the late 1970s, blast furnace slag was tipped over most of the site, raising the ground to 5 m–7 m AOD to allow access. Ground levels needed to be further raised, typically by about 1 m, to 5 m–8 m AOD. An appropriate form of ground treatment was needed that would minimise post-construction settlements of the development infrastructure (road/services corridors) and enable the development programme to be met.

Deficiencies in ground behaviour (Section 3)

The site comprised a sequence of highly compressible clays, silts and peat up to 25 m deep. Previous filling on the site indicated that the additional landfilling would exceed any pre-consolidation pressures within the compressible deposits, thus causing large primary consolidation settlements. Furthermore, the peat layer would give rise to large long-term secondary compression settlements. A geological cross-section is presented in Figure A.3.1.

Engineering properties (Section 4)

The results of laboratory tests performed on 100 mm diameter open tube samples and piston samples are summarised in Table A.3.1. *In situ* test results and permeability data obtained by falling and rising head tests in standpipe piezometers are also given.

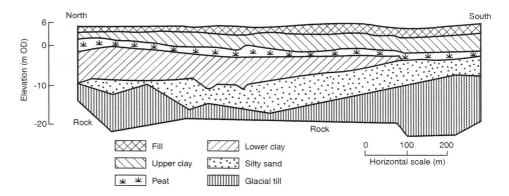

Figure A.3.1 *Geological cross-section (after Jones et al, 1995)*

Table A.3.1 *Summary of engineering properties*

Properties	Desiccated clay	Upper silty clay	Peat	Lower silty clay	Silty sand lenses	Glacial till/ gravels
Moisture content (%)	30	52	419	45	35	11
Liquid limit (%)	50	52	589	47	32	28
Plastic limit (%)	25	24	408	24	22	17
Clay/silt fraction (%)	55	36	–	37	33	18
c_u (kPa): laboratory	32	19	32	21	40	216
in situ vane	18	22	37	24	–	–
m_v (m_2/MN)	0.38	0.94	2.17	0.52	0.21	–
c_v (m^2/yr)	1.2	1.44	>6.4	1.84	10.8	–
C_α	–	0.016	0.037	0.007	–	–
c_r (m^2/yr) swelling	1.03	0.52	0.11	1.88	14.6	–
Permeability k_h (m/sec)	–	4×10^{-9}	3×10^{-8}	9×10^{-9}	–	–

Measurement of engineering properties (Section 5)

Two trial surcharges were constructed on different parts of the site to enable consolidation times to be evaluated and a trial excavation in the existing slag fill was made to investigate unloading behaviour. Instrumentation included high air entry piezometers and magnet settlement gauges. The results are shown in Table A.3.2.

Provision of ground treatment (Section 6 and 8.2)

A geotechnical model was developed from the laboratory and trial data, which assumed that loading the ground would produce primary consolidation and large settlements. All subsequent settlement of the roadways should, therefore, be related to secondary compression. The width of the road corridor including service reserves was 12 m. Calculations showed that the optimum surcharge design comprised an embankment with a 25 m crest width, 1 in 4 side slopes and a 4.6 m maximum height.

Table A.3.2 *Properties derived from pre-loading and excavation trials*

Properties	Upper silty clay	Peat	Lower silty clay
m_v (m²/MN)	0.71	3.42	0.55
m_s (m²/MN) swelling	0.24	0.33	0.07
c_v (m²/yr)	62	–	75
C_α	0.01	0.028	0.007
c_r (m²/yr) swelling	10	–	101
Permeability k_h (m/sec)	1.4×10^{-8}	–	1.3×10^{-8}

The criterion adopted for the removal of the surcharge was that the degree of consolidation at the centre of the compressible deposits should be at least equivalent to the proposed landfill loading. The time required for the surcharge to be left in place was estimated to be of the order of six months for each phase of the operation.

Following removal of the surcharge and road construction, permanent movement pins were set in kerblines. Representative actual settlements were monitored over a five-year period and plotted against the logarithm of time and the results extrapolated to predict 25-year settlements. In a surcharged area the greatest predicted settlement was less than 150 mm and predicted settlements were generally less than 60 mm. By comparison, in a limited area at chainage 150, where surcharge was not possible, much larger movements occurred despite only 600 mm of landfill being placed. The results plotted on Figure A.3.2 show that the magnitude of the predicted settlements based on actual observations was substantially reduced by the surcharge.

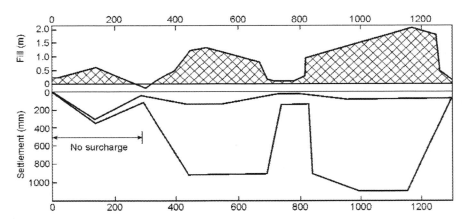

Figure A.3.2 *Observed and predicted settlements (after Jones et al., 1995)*

MIXED CLAY FILL – VIBRO STONE COLUMNS

Reference: Watts and Charles (1991)

Commentary

There are significant variations in pre-treatment ground conditions at this site, both with depth and in lateral extent. The soft clay fill in the northern half of the site (where the skip load tests were carried out) was probably marginal for the application of stone columns, in particular using the dry top-feed process. Lateral resistance in this soil may not have been sufficient to prevent excessive bulging of the columns under load. Variability of ground conditions was not reflected in the design of vibro to support the building straddling both halves of the site. The flexible nature of the building during the initial construction phase, when a high proportion of the total load was applied, probably prevented any serviceability problems with the more rigid infill panels or internal finishings.

Required ground behaviour (Section 2)

The development consists of a two-storey steel frame structure with concrete blockwork infill panels. The principal source of ground movement was anticipated to be from structural loading. The problems to be addressed therefore were the provision of adequate bearing capacity and limiting the differential settlement.

Deficiencies in ground behaviour (Section 3)

A former gravel pit at Abingdon was backfilled with miscellaneous clay fill in the 1960s. For a number of years the southern half of the site was used as a storage compound for heavy civil engineering plant, while the northern part had been grassed over and had remained undisturbed since backfilling. The remedial strategy was to treat all the ground.

Engineering properties (Section 4)

The building has one wing on the southern half of the site and the other on the northern half. The two wings are joined across their western ends by a similar structure. All structural columns are supported on reinforced concrete foundation pads linked by reinforced concrete ground beams to support the blockwork infill.

Two trial pits in the southern part of the site showed tarmac, hardcore fill and lean mix concrete in the top 1.0 m, underlain by 0.6 m of firm clay fill. Beneath this were 2.2 m of soft clay fill and 0.5 m of sandy gravel. In the northern area, two trial pits revealed a firm clay fill up to 1.6 m thick overlying 2.0 m of soft clay and silt fill. This fill contained organic material. The bottom of the original excavation was at 4.3 m depth and was a firm natural clay deposit.

Measurement of engineering properties (Section 5)

Two full-scale load tests were carried out on fill close to the north wing as shown in Figure A.4.1. A group of four stone columns was constructed during ground treatment and a 2.0 m square pad was cast over them to model a foundation pad. A 2.0 m square concrete pad was also cast on untreated ground. Each pad was loaded with two sand filled skips and settlement was measured over a six-month period.

Figure A.4.1 *Skip load tests (courtesy Building Research Establishment Ltd)*

The two skips applied an additional 50 kPa to the fill. This represented 60 per cent of the estimated structural loading imposed on foundation pads 1, 2, 5 and 6 of the structure, see Figure A.4.2. The properties of the stiff and softer clay fill were measured in the laboratory on samples obtained from boreholes adjacent to each test and estimates of the settlement of each pad had no ground treatment been carried out were computed. The load test on untreated ground settled 13 mm while the pad on the four stone columns settled 18 mm. Predictions from laboratory tests for untreated ground gave similar results. This suggests that the columns constructed under one of the pads had little effect in reducing its total settlement.

Provision of ground treatment (Section 6 and 9.2)

Prior to construction, the fill was treated using the dry top-feed vibro technique. Stone columns were constructed through the full depth of the fill. Depending on structural loading, one, two, three or four columns were placed at the location of each foundation pad with single rows of columns under the ground beams. A few treatment points were positioned under the ground floor slabs. The foundation plan layout and location of treatment points is shown in Figure A.4.2.

Monitoring of the settlement of the foundations on the north and south wings began before the steel frame was erected. Levelling was transferred to adjacent blockwork panels when the foundations were covered. The average settlements of each wing of the building over a fourteen month period from the beginning of construction is plotted in Figure A.4. 3. During that period the monitored part of the south wing has settled an average of 7 mm. The seven points monitored on the north wing over the same period settled an average settlement of 15 mm. Measurement of the settlement of the building has continued for 12 years following construction.

Over this period the south wing of the building has settled about half as much as the north wing, with an average settlement of 18 mm for pads 1, 2, 5 and 6 in the north wing. It would seem that vibro has not eliminated significant differential settlements in the variable fill across the site. However, the total movements are relatively small and a major proportion of the settlement was built out during construction of the flexible steel frame building. There is no evidence to date of any distress to the completed structure.

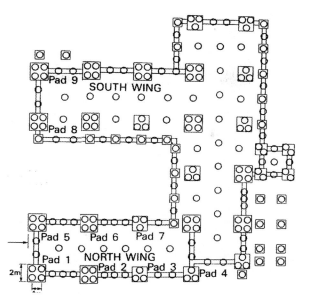

Figure A.4.2 *Layout of vibro stone columns at Abingdon (after Watts and Charles, 1991)*

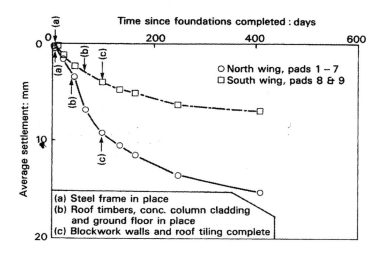

Figure A.4.3 *Settlement of the building at Abingdon (after Watts and Charles, 1991)*

References: Wood *et al* (1996)

                   Watts *et al* (2000)

Commentary

This case history demonstrates the role of testing treated ground in understanding ground treatment.

- The benefits of the vibro treatment are evident although it is apparent that the effect of the stone columns in significantly limiting foundation settlement only occurred at higher loads. The maximum settlement of the foundation was reduced by 40 per cent by the installation of stone columns.

- An important objective of vibro is to reduce differential movements, particularly in variable ground conditions. There was significantly less total settlement of the treated strip where the fill is deepest and consequently smaller differential settlement in the treated strip foundation

- As well as reinforcing poor soils, an objective of vibro in coarse soils is to improve the stiffness of the soil through radial densification. This trial clearly demonstrated the potential for *in situ* densification in a miscellaneous fill but also showed the important role of workmanship, the appropriate choice of plant and treatment process.

Required ground behaviour (Section 2)

It is difficult to predict the behaviour of foundations on variable fills in which vibro stone columns have been installed. The interaction of the modified fill, stone columns and the foundation is complex. Full scale instrumented load tests were carried out at an experimental site in Bacup to study the installation and performance of vibro stone columns supporting a strip foundation in a variable fill and the performance of a similar strip foundation on untreated ground. The objective of ground treatment was to reduce total and differential settlement and the objective of the tests was to gain a better understanding of the design and application of vibro in very variable fills.

Deficiencies in ground behaviour (Section 3)

An initial site investigation comprising five trial pits and two boreholes identified granular fill overlying cohesive fill. There was stiff glacial till below the fill. Samples obtained from boreholes drilled for instrumentation showed the granular fill material was very variable. It contained a high proportion of black ash, some small pieces of sandstone and limestone, slate, burnt slag, clinker, brick and concrete fragments, sand and gravel and sandstone cobbles and boulders. The coarse fill contained pockets of soft silty clay. The lower cohesive fill comprised silty clay with dispersed granular fragments. The site investigation data from the immediate vicinity of the treated and untreated test foundations are summarised in Figure A.5.1 and illustrate the considerable variability of the fill and the potential for unacceptable differential settlements of a foundation on this fill.

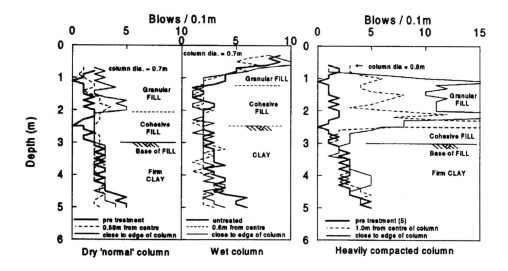

Figure A.5.2 *Dynamic probing around vibro stone columns (after Watts et al, 2000)*

In addition to surface movements, settlement was measured at depth within the fill and the underlying firm natural clay deposit beneath both treated and untreated strips. They showed that the presence of the columns gave rise to more deep-seated settlements than beneath the untreated strip. These settlements occurred within the cohesive fill and the underlying stiff clay below the base of the columns. It appears that the presence of the columns encouraged load transfer to a greater depth resulting in the deeper seated settlements.

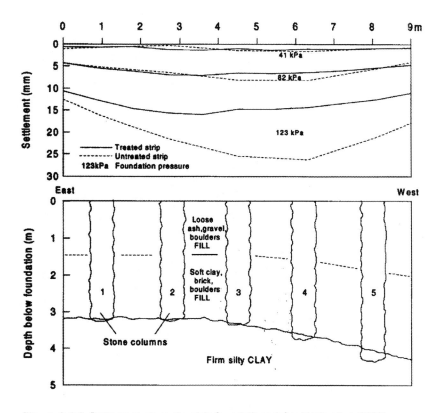

Figure A.5.3 *Settlement along the strip foundations (after Watts et al, 2000)*

HIGHLY COMPRESSIBLE SOILS – VIBRO CONCRETE COLUMNS

Reference: Maddison *et al* (1996)

Commentary

This case history describes the first major application of vibro concrete columns in the United Kingdom. The design incorporated a load transfer platform reinforced with geogrids, which supported the embankment between columns, transferring the load to the columns and hence to the underlying stable stratum. This approach has now become relatively common, both for vibro concrete columns supporting widespread loads and similar applications of vibro stone columns. Zone tests modelled the proposed loading in detail and included a high level of instrumentation.

- The zone tests showed that the reinforced platform could span between columns to a sufficient degree to minimise differential settlement across the reinforced soil mass. Earth pressure measurements confirmed that load transfer was taking place.

- After some initial settlement under load, sufficient end-bearing resistance was developed in the VCCs and very little additional movements took place.

Required ground behaviour (Section 2)

The new toll plaza for the Second Severn Crossing was constructed on low lying land adjacent to the estuary. An embankment, generally 2.5 m to 3.5 m high with a maximum height of some 6.0 m, was constructed over highly compressible peat and clay soils.

The principal problems to be addressed were excessive and prolonged primary consolidation, large long-term secondary compression settlements over the design life of the structure and differential movements along the carriageway. An innovative foundation system, comprising vibro concrete columns and a load transfer platform incorporating low strength geogrids at the base of the embankment, was chosen to support the embankment.

Deficiencies in ground behaviour (Section 3)

For embankment construction of up to 3.5 m height using conventional fill, consolidation settlements as large as 0.65 m were anticipated in the founding estuarine clay and peat soils. At the west end of the toll plaza, maximum settlements of 1.0 m were calculated. It was estimated that primary consolidation of these deposits might take between six months and three years. In addition secondary compression settlements of 0.2 m over a 120-year design life were calculated. Design criteria of 100 mm maximum post-construction settlement over 120 years with an angular distortion due to differential movements not exceeding 1 in 1000 were adopted.

Engineering properties (Section 4)

The ground investigation included 11 boreholes sunk by cable percussion and rotary coring techniques, 40 cone penetration tests and comprehensive laboratory testing. A summary of the properties is presented in Table A.6.1

Table A.6.1 *Summary of engineering properties*

Design parameters	Desiccated clay	Estuarine clay	Peat	Sands and gravels
Moisture content (%)	24	53	215	14
Liquid limit (%)	52	54	262	NA
Plastic limit (%)	24	25	164	NA
Plasticity index (%)	29	28	98	NA
Organic content (%)	8	5	46	NA
Undrained shear strength: c_u (kPa)	40	15	17	NA
Effective shear strength: c' (kPa), ϕ' (degrees)	5, 25	0, 29	10, 23	0, 30
Coefficient of volume compressibilty: m_v (m²/MN)	0.20	0.57	2.00	NA
Coeff. Of consolidation: c_v (m²/yr)	2.5	2.0	12	NA
Coeff. Of secondary compression: C_α	0.002	0.004	0.027	NA
Deformation modulus (MPa)	–	–	–	62[1]
Shear modulus (MPa)	–	–	–	27[1]

Notes :   NA – Not applicable

(1) – Derived from back-analysis of VCC load tests

Measurement of engineering properties (Section 5)

Site tests included static and dynamic load testing of two VCCs and a zone test modelling the proposed permanent works design in an area where ground conditions were considered typical. The results of the static load tests on individual VCCs are summarised in Figure A.6.1.

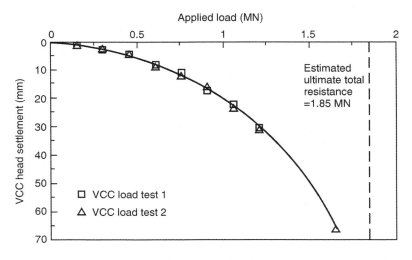

Figure A.6.1 *VCC load test results (after Maddison* et al, *1996)*

The zone test covered an area of 7.2 m × 7.2 m, at the centre of which four VCCs were installed on a 2.5 m square grid. A load transfer platform was constructed at the ground surface and loading of 60 kPa was applied to model embankment loading. The test and results are shown in Figures A.6.2 and A.6.3 respectively.

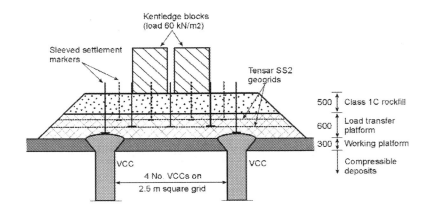

Figure A.6.2 *VCC zone test (after Maddison et al, 1996)*

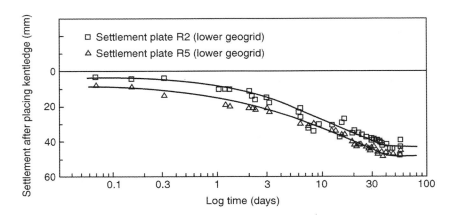

Figure A.6.3 *VCC zone test results (after Maddison et al, 1996)*

Provision of ground treatment (Section 6 and 9.5)

Based on the ground investigation data and the results of the VCC load tests, and employing an overall factor of safety of 3.0, a design load of 0.6 MN maximum was adopted for individual VCCs. The spacing between VCCs was typically 2.7 m on triangular grid but reduced with increasing embankment height to 2.2 m. About 11700 VCCs were installed over a five-month period. The columns were typically 5 m to 6 m in length and penetrated the founding sand and gravel soils to 0.5–1.0 m depth and a bulbous base was formed in these deposits. The VCCs were formed with an expanded head profile. Construction of the load transfer platform followed the installation of the VCCs.

Settlement observations for construction up to final earthworks levels indicated 5 mm to 10 mm above the VCCs and 8 mm to 15 mm at mid span between them. Placing fill to model the road construction and traffic loading caused settlements to increase to 15 mm–35 mm above the VCCs and 19 mm and 42 mm at mid span. The differential settlement in the load transfer platform between VCCs and mid span was generally less than 5 mm. It is suggested that the increased settlement was due to displacement of the VCCs into the founding stratum consistent with the development of end bearing resistance. Where VCCs were founded in silty sand, settlements of 35 mm to 75 mm occurred. These larger settlements would seem to be the consequence of the lower bearing capacity of the soils in that location and reduced penetration of the VCCs into founding soils. Such movements were accommodated given the flexible nature of the load transfer platform and the embankment.

Level surveys over the 12 months following the completion of road construction revealed no further movements of the embankment supported on VCCs and the load transfer platform.

References

ABBISS C P (1981)
Shear wave measurements of the elasticity of the ground, *Géotechnique*, vol 31, no 1, pp 91–104

AHNBERG H, BENGTSSON P-E and HOLM G (1989)
Prediction of strength of lime columns, in *Proceedings of 12th International Conference on Soil Mechanics and Foundation Engineering*, Rio de Janeiro, vol 2, pp 1327–1330 (Balkema, Rotterdam)

AHNBERG H and HOLM G (1999)
Stabilisation of some Swedish organic soils with different types of binder, in *Dry mix methods for deep soil stabilisation* (Bredenberg H, Holm G and Broms B B, eds), Proceedings of International Conference, Stockholm, pp 101–108 (Balkema, Rotterdam)

AMERICAN SOCIETY OF CIVIL ENGINEERS (1978)
Soil improvement, history, capabilities and outlook, Report by committee on placement and improvement of soils of Geotechnical Engineering Division (ASCE, New York)

ANDRUS R D, CHUNG R M, STOKOE K H and BAY J A (1998)
Delineation of densified sand at Treasure Island by SASW testing, in *Geotechnical site characterisation* (Robertson P K and Mayne P W eds), *Proceedings of 1st International Conference*, Atlanta, vol 1, pp 459–464 (Balkema, Rotterdam)

ANON (1979)
Dynamic consolidation at Surrey docks, *Ground Engineering*, vol 12, no 1, pp 32–34

ANON (1997)
Window shopping – Ground Engineering looks at the rise of the window sampler in UK site investigations, *Ground Engineering*, vol 30, no 9, October, p18

ANON (1999)
Irlam rail embankment, *Ingenia:* Informative Quarterly of Royal Academy of Engineering, vol 1, no 2, October, pp 41–42

ASSOCIATION OF GEOTECHNICAL AND GEOENVIRONMENTAL SPECIALISTS (1998a)
Code of conduct for site investigation (AGS, Beckenham, Kent)

ASSOCIATION OF GEOTECHNICAL AND GEOENVIRONMENTAL SPECIALISTS (1998b)
Guidelines for good practice in site invstigation (AGS, Beckenham, Kent)

ASSOCIATION OF GEOTECHNICAL AND GEOENVIRONMENTAL SPECIALISTS (1998c)
AGS guide: the selection of geotechnical soil laboratory testing (AGS, Beckenham, Kent)

ATKINSON M F (1993)
Structural foundations manual for low-rise buildings (Spon, London)

BALAAM N P and BOOKER J R (1981)
Analysis of rigid rafts supported by granular piles. *International Journal for Numerical and Analytical Methods in Geomechanics*, vol 5, no 4, pp 379–403

BARRON R A (1948)
Consolidation of fine-grained soils by drain wells, *ASCE Transactions*, vol 113, pp 718–754

BAUMANN V and BAUER G E A (1974)
The performance of foundations on various soils stabilised by the vibro-compaction method. *Canadian Geotechnical Journal*, vol 11, pp 509–530

BELL A L, SLOCOMBE B C, NESBITT A M and FINEY J T (1986)
Vibro-compaction densification of a deep hydraulic fill, in *Building on marginal and derelict land*, Proceedings of Institution of Civil Engineers Conference, Glasgow, May 1986, pp 791–797 (Thomas Telford, London, 1987)

BELL F G and CASHMAN P M (1985)
Ground water control by ground water lowering, in *Groundwater in engineering geology* (Cripps J C, Bell F G and Culshaw M G, eds), Proceedings of 21st Annual Conference of Engineering Group of Geological Society, Sheffield, September 1985, pp 471–486 (Geological Society, London, 1986)

BELL S E (1977)
Successful design for mining subsidence, in *Large ground movements and structures* (Geddes J D, ed), Proceedings of Conference, Cardiff, July 1977, pp 562–578 (Pentech Press, Plymouth, 1978)

BIELBY S C (2001)
Site safety handbook (third edition) Special Publication 151 (CIRIA, London)

BJERRUM L (1973)
General report on problems of soil mechanics and construction on soft clays and structurally unstable soils (collapsible, expansive and others), in *Proceedings of 8th International Conference on Soil Mechanics and Foundation Engineering*, Moscow, vol 2.3, pp 111–159

BOONE S J (1996)
Ground movement related building damage, *ASCE Journal of Geotechnical Engineering*, vol 122, no 11, November, pp 886–896

BOSCARDIN M D and CORDING E J (1989)
Building response to excavation-induced settlement, *ASCE Journal of Geotechnical Engineering*, vol 115, no 1, pp 1–21

BREDENBERG H (1983) Lime columns for ground improvement at new cargo terminal in Stockholm, in *Improvement of ground* (Rathmayer H G and Saari K H O, eds), Proceedings of 8th European Conference on Soil Mechanics and Foundation Engineering, Helsinki, vol 2, pp 881–884 (Balkema, Rotterdam)

BREDENBERG H and BROMS B B (1983)
Lime columns as foundations for buildings, in *Piling and ground treatment*, Proceedings of International Conference, London, March 1983, pp 133–138 (Thomas Telford, London, 1984)

BRIAUD J-L, LIU M L and LEPERT P (1989)
The WAK test to check the increase in soil stiffness due to dynamic compaction, in *Geotechnics of waste fills: theory and practice*, Proceedings of Conference, Pittsburg, September 1989, pp 107–122, Special Technical Publication 1070 (American Society for Testing and Materials, Philadelphia, 1990)

BRIAUD J-L and LEPERT P (1990)
WAK test to find spread footing stiffness, *ASCE Journal of Geotechnical Engineering*, vol 116, no 3, March, pp 415–431

BRIERLEY G S (1998) Subsurface investigations and geotechnical report preparation, in *Subsurface conditions – risk management for design and construction management professionals* (Hatem D J, ed), pp 49–94 (Wiley, New York)

BRITISH STANDARDS INSTITUTION (1986)
Code of practice for foundations, BS8004:1986 (British Standards Institution, London)

BRITISH STANDARDS INSTITUTION (2001)
Investigation of potentially contaminated sites – Code of practice BS 10175:2001, (British Standards Institution, London)

BRITISH STANDARDS INSTITUTION (1990)
Methods of test for soils for civil engineering purposes (9 parts), BS1377: 1990 (British Standards Institution, London)

BRITISH STANDARDS INSTITUTION (1993)
Flat-bottomed, vertical, cylindrical storage tanks for low temperature service: Part 3 - Recommendations for the design and construction of prestressed and reinforced concrete tanks and tank foundations, and for the design and installation of tank insulation, tank liners and tank coatings BS 7777: Part 3: 1993 (British Standards Institution, London)

BRITISH STANDARDS INSTITUTION (1995)
Eurocode 7: Geotechnical design – Part 1: General rules (Draft for development) DD ENV 1997-1: 1995 (British Standards Institution, London)

BRITISH STANDARDS INSTITUTION (1999)
Code of practice for site investigations, BS5930: 1999, (British Standards Institution, London)

BRITISH STANDARDS INSTITUTION (2000a)
Eurocode 7: Geotechnical design – Part 2: Design assisted by laboratory testing (Draft for development) DD ENV 1997-2: 2000 (British Standards Institution, London)

BRITISH STANDARDS INSTITUTION (2000b)
Eurocode 7: Geotechnical design – Part 3: Design assisted by field testing (Draft for development) DD ENV 1997-3: 2000 (British Standards Institution, London)

BROMS B B (1993)
Lime stabilisation, in *Ground improvement* (M P Moseley, ed), pp 65–99 (Blackie, Glasgow)

BROMS B B (1999)
Keynote lecture: Design of lime, lime/cement and cement columns, in *Dry mix methods for deep soil stabilisation* (Bredenberg H, Holm G and Broms B B, eds), Proceedings of International Conference, Stockholm, pp 125–153 (Balkema, Rotterdam)

BRUCE D A, BRUCE M E C and DIMILLIO A F (1999)
Dry Mix Methods: A brief overview of international practice, in *Dry mix methods for deep soil stabilisation* (Bredenberg H, Holm G and Broms B B, eds), Proceedings of International Conference, Stockholm, pp 15–25 (Balkema, Rotterdam)

BUILDING RESEARCH ESTABLISHMENT (1987a)
Site investigation for low-rise building: desk studies, Digest 318 (BRE, Garston)

BUILDING RESEARCH ESTABLISHMENT (1987b)
Site investigation for low-rise building: procurement, Digest 322 (BRE, Garston)

BUILDING RESEARCH ESTABLISHMENT (1989a)
Site investigation for low-rise building: the walk-over survey, Digest 348 (BRE, Garston)

BUILDING RESEARCH ESTABLISHMENT (1989b)
Simple measuring and monitoring of movements in low-rise buildings – Part 1: cracks, Digest 343 (BRE, Garston)

BUILDING RESEARCH ESTABLISHMENT (1989c)
Simple measuring and monitoring of movements in low-rise buildings – Part 2: settlement, heav and out-of-plumb, Digest 344 (BRE, Garston)

BUILDING RESEARCH ESTABLISHMENT (1993a)
Site investigation for low-rise building: trial pits, Digest 381 (BRE, Garston)

BUILDING RESEARCH ESTABLISHMENT (1993b)
Site investigation for low-rise building: soil description, Digest 383 (BRE, Garston)

BUILDING RESEARCH ESTABLISHMENT (1993c)
Low-rise buildings on shrinkable clay soils: Part 1, Digest 240 (BRE, Garston)

BUILDING RESEARCH ESTABLISHMENT (1995)
Site investigation for low-rise building: direct investigations, Digest 411, (BRE, Garston)

BUILDING RESEARCH ESTABLISHMENT (1997) *Low-rise buildings on fill: classification and load-carrying characteristics*, Digest 427: Part 1 (BRE, Garston)

BUILDING RESEARCH ESTABLISHMENT (1998a) *Low-rise buildings on fill: site investigation, ground movement and foundation design*, Digest 427: Part 2 (BRE, Garston)

BUILDING RESEARCH ESTABLISHMENT (1998b)
Low-rise buildings on fill: engineered fill, Digest 427 Part 3 (BRE, Garston)

BUILDING RESEARCH ESTABLISHMENT (2000)
Specifying vibro stone columns (Construction Research Communications, London)

BURFORD D (1991)
Surcharging a deep opencast backfill for housing development, *Ground Engineering*, September, pp 36–39

BURFORD D and CHARLES J A (1991)
Long term performance of houses built on opencast ironstone mining backfill at Corby, 1975–1990, in *Ground movements and structures* (Geddes J D, ed), Proceedings of 4th International Conference, Cardiff, July 1991, pp 54–67 (Pentech Press, London, 1992)

BURLAND J B and WROTH C P (1974)
Settlement of buildings and associated damage: review paper, in *Settlement of structures*, Proceedings of British Geotechnical Society Conference, Cambridge, 1974, pp 611–654 (Pentech Press, London, 1975)

BUTCHER A P and McELMEEL K (1993)
The ability of *in situ* testing to assess ground treatment, in *Engineered fills*, Proceedings of International Conference, Newcastle-upon-Tyne, September, pp 529–540 (Thomas Telford, London)

BUTCHER A P, McELMEEL K and POWELL J J M (1995)
Dynamic probing and its use in clay soils, in *Advances in site investigation practice*, Proceedings of International Conference, London, March, pp 383–395 (Thomas Telford, London)

CASAGRANDE L (1947)
The application of electro-osmosis to practical problems in foundations and earthworks, Building Research Technical Paper no 30 (HMSO, London)

CHANG J C E (1981)
Long-term consolidation beneath the test fills at Vasby, Sweden, Swedish Geotechnical Institute, Report no 13

CHARLES J A (1993)
Building on fill: geotechnical aspects, Building Research Establishment Report BR230 (Garston, Watford)

CHARLES J A (1996)
The depth of influence of loaded areas, *Geotechnique*, vol 46, no 1, pp 51–61

CHARLES J A, NAISMITH W A and BURFORD D (1977)
Settlement of backfill at Horsley restored opencast coal mining site, in *Large ground movements and structures* (Geddes J D, ed), Proceedings of Conference, Cardiff, July 1977, pp 229–251 (Pentech Press, Plymouth, 1978)

CHARLES J A, EARLE E W and BURFORD D (1978)
Treatment and subsequent performance of cohesive fill left by opencast ironstone mining at Snatchill experimental housing site, Corby, in *Clay fills*, Proceedings of Conference, London, November 1978, pp 63–72 (Institution of Civil Engineers, London, 1979)

CHARLES J A and DRISCOLL R (1981)
A simple *in situ* load test for shallow fill, *Ground Engineering*, vol 14, no 1, pp 31–36

CHARLES J A, BURFORD D and WATTS K S (1981)
Field studies of the effectiveness of "dynamic consolidation", in *Proceedings of 10th International Conference on Soil Mechanics and Foundation Engineering*, Stockholm, vol 3, pp 617–622

CHARLES J A and WATTS K S (1982)
A field study of the use of the dynamic consolidation ground treatment technique on soft alluvial soil, *Ground Engineering*, vol 15, no 5, pp 17–22 and 25

CHARLES J A and WATTS K S (1983)
Compressibility of soft clay reinforced with granular columns, in *Improvement of ground* (Rathmayer H G and Saari K O, eds), Proceedings of 8th European Conference on Soil Mechanics and Foundation Engineering, Helsinki, vol 1, pp 347–352 (Balkema, Rotterdam)

CHARLES J A, BURFORD D and WATTS K S (1986)
Improving the load carrying characteristics of uncompacted fills by preloading, *Municipal Engineer*, vol 3, no 1, pp 1–19

CHARLES J A and BURFORD D (1987)
Settlement and groundwater in opencast mining backfills, in *Proceedings of 9th European Conference on Soil Mechanics and Foundation Engineering*, Dublin, vol 1, pp 289–292

CHARLES J A, BURFORD D and HUGHES D B (1993)
Settlement of opencast mining backfill at Horsley 1973–1992, in *Engineered fills* (Clarke B G, Jones C J F P and Moffat A I B, eds), Proceedings of International Conference, Newcastle-upon-Tyne, pp 429–440 (Thomas Telford, London)

CHARLES J A and WATTS K S (1996)
The assessment of the collapse potential of fills and its significance for building on fill, *Geotechnical Engineering*, Proceedings of Institution of Civil Engineers, vol 119, January, pp 15–28

CHARLES J A, SKINNER H D and WATTS K S (1998)
The specification of fills to support buildings on shallow foundations: the "95 per cent fixation", *Ground Engineering*, vol 31, no 1, January, pp 29–33

CHOA V (1994)
Application of the observational method to hydraulic fill reclamation projects, *Géotechnique*, vol 44, no 4, pp 735–745

CHOA V, KARANARATNE G P, RAMASWAMY S D, VIJIARATNAM A and LEE S L (1979)
Compaction of sand fill at Changi airport, in *Proceedings of 6th Asian Regional Conference on Soil Mechanics and Foundation Engineering*, Singapore, vol 1, pp 137–140

CHOW Y K, YONG D M, YONG K Y and LEE S L (2000)
Improvement of granular soils by high-energy impact, *Ground Improvement*, vol 4, no 1, January, pp 31–35

CLARK R G (1998)
Costs that could have been saved by a desk study: a case history, in *The value of geotechnics in construction*, Proceedings of Seminar held at Institution of Civil Engineers, pp 65–72 (CRC, London)

CLARKE B G (1995)
Pressuremeters in geotechnical design (Blackie, London)

CLARKE B G, JONES C J F P and MOFFAT A I B (eds) (1993)
Engineered fills, Proceedings of Conference, Newcastle upon Tyne (Thomas Telford, London)

CLAYTON C R I (1993)
The standard penetration test (SPT): methods and use, CIRIA Report 143 (CIRIA, London)

CLAYTON C R I, MATTHEWS M C and SIMONS N E (1995)
Site investigation (Blackwell Science, Oxford)

CODUTO D P (1999)
Geotechnical engineering – principles and practices (Prentice Hall, New Jersey)

COOPER R and ROSE A N (1999)
Stone column support for an embankment on deep alluvial soils, *Geotechnical Engineering*, Proceedings of Institution of Civil Engineers, vol 137, January, pp 15–25

COVIL C S, LUK MW and PICKLES A R (1997)
Case history: ground treatment of the sand fill at the new airport at Chep Lai Kok, Hong Kong, in *Ground improvement geosystems: densification and reinforcement* (Davies M C R and Schlosser F, eds), Proceedings of 3rd International Conference, London, pp 149–156 (Thomas Telford, London)

CRAIG W H and AL-KHAFAJI Z A (1997)
Reduction of soft clay settlement by compacted sand columns, in *Ground improvement geosystems: densification and reinforcement* (Davies M C R and Schlosser F, eds), Proceedings of 3rd International Conference, London, pp 218–231 (Thomas Telford, London)

DEGEN W S (1997) 56 m deep vibrocompaction at German lignite mining area, in *Ground improvement geosystems: densification and reinforcement* (Davies M C R and Schlosser F, eds), Proceedings of 3rd International Conference, London, pp 127–133 (Thomas Telford, London)

DOWNIE A R and TREHARNE G (1979)
Dynamic consolidation of refuse at Cwmbran, in *Engineering behaviour of industrial and urban fill*, Proceedings of Symposium, Birmingham, pp E15–E24 (Midland Geotechnical Society, Birmingham)

DUDLEY J H (1970)
Review of collapsing soils, *ASCE Journal of Soil Mechanics and Foundations Division*, vol 96, no SM3, pp 925–947

DUNNICLIFF J (1988)
Geotechnical instrumentation for monitoring field performance (Wiley, New York)

EAKIN W R G and CROWTHER J (1985)
Geotechnical problems on land reclamation sites, *Municipal Engineer*, vol 2, October, pp 233–245

ECCLES C S AND REDFORD R P (1997)
The use of dynamic (window) sampling in the site investigation of potentially contaminated ground, in *Geoenvironmental engineering – Contaminated ground: fate of pollutants and remediation* (Yong R N and Thomas H R, eds), Proceedings of British Geotechnical Society Conference, Cardiff, pp 11–16 (Thomas Telford, London)

ERIKSSON L and EKSTOM A (1983)
The efficiency of three different types of vertical drain – results from a full-scale test, in *Improvement of ground* (Rathmayer H G and Saari K O, eds), Proceedings of 8th European Conference on Soil Mechanics and Foundation Engineering, Helsinki, vol 2, pp 605–610 (Balkema, Rotterdam)

ESRIG M I (1999)
Keynote lecture: Properties of binders and stabilised soils, in *Dry mix methods for deep soil stabilisation* (Bredenberg H, Holm G and Broms B B, eds), Proceedings of International Conference, Stockholm, pp 67–72 (Balkema, Rotterdam)

FRASER R A and WARDLE L J (1976)
Numerical analysis of rectangular rafts on layered foundations, *Geotechnique*, vol 26, no 4, pp 613–630

GANJI V, GUCUNSKI N and MAHER A (1997)
Detection of underground obstacles by SASW methods – numerical aspects, *ASCE Journal of Geotechnical and Geoenvironmental Engineering*, vol 123, no 3, March, pp 212–219

GLENDINNING S and ROGERS C D F (1996)
Deep stabilisation using lime, in *Lime stabilisation* (Rogers C D F, Glendinning S and Dixon N, eds), Proceedings of Seminar held at Loughborough University, pp 127–138 (Thomas Telford, London)

GREENWOOD D A and KIRSCH K (1983)
Specialist ground treatment by vibratory and dynamic methods, in *Piling and ground treatment*, Proceedings of International Conference, London, March 1983, pp 17–45 (Thomas Telford, London, 1984)

HAEGEMAN W and VAN IMPE W F (1998)
SASW control of a vacuum consolidation on a sludge disposal, in *Geotechnical site characterisation* (P K Robertson and P W Mayne, eds), Proceedings of 1st International Conference, Atlanta, vol 1, pp 473–477 (Balkema, Rotterdam)

HALKOLA H (1999)
Keynote lecture: Quality control for dry mix methods, in *Dry mix methods for deep soil stabilisation* (Bredenberg H, Holm G and Broms B B, eds), Proceedings of International Conference, Stockholm, pp 285–294 (Balkema, Rotterdam)

HANSBO S (1993)
Band drains, in *Ground improvement* (Moseley M P, ed), pp 40–64 (Blackie, Glasgow)

HATEM D J (1998)
Introduction, in *Subsurface Conditions – risk management for design and construction management professionals* (Hatem D J, ed), pp 1–4 (Wiley, New York)

HAWKINS A B (ed) (1997)
Ground chemistry implications for construction (Balkema, Rotterdam)

HEAD K H (1998)
Manual of soil laboratory testing (3 vols), second edition (Wiley, Chichester)

HEALTH and SAFETY COMMISSION (1992)
Management of Health and Safety at Work Regulations, Approved Code of Practice (HMSO, London)

HEALTH and SAFETY COMMISSION (1994) *Managing construction for health and safety, Construction (Design and Management) Regulations (1994), Approved Code of Practice L54* (HSE Books)

HEATHCOTE F W L (1965)
Movement of articulated buildings on subsidence sites, *Proceedings of Institution of Civil Engineers*, vol 30, pp 347–368

HENDY M S and MUIR I C (1997)
Experience of dynamic replacement on a 40 m deep reclamation in Hong Kong, in *Ground improvement geosystems: densification and reinforcement* (Davies M C R and Schlosser F, eds), Proceedings of 3rd International Conference, London, pp 76–81 (Thomas Telford, London)

HOLTZ R D, JAMIOLKOWSKI M B, LANCELLOTTA R AND PEDRONI R (1991)
Prefabricated vertical drains: design and performance (CIRIA Ground Engineering Report, Butterworth-Heinemann, Oxford)

HORIKOSHI K and RANDOLPH M F (1997)
On the definition of raft-soil stiffness ratio for rectangular rafts, *Geotechnique*, vol 47, no 5, pp 1055–1061

HUGHES J M O and WITHERS N J (1974)
Reinforcing of soft cohesive soils with stone columns, *Ground Engineering*, vol 7, no 3, May, pp 42–49

HUMPHESON C, SIMPSON B and CHARLES J A (1991)
Investigation of hydraulically placed pfa as a foundation for buildings, in *Ground movements and structures* (Geddes J D, ed), Proceedings of 4th International Conference, Cardiff, July 1991, pp 68–88 (Pentech Press, London, 1992)

ILANDER A, FORSMAN J and LAHTINEN P (1999)
Combined mass and column stabilisation in Kivikko test embankment – designing by traditional and FE-methods, in *Dry mix methods for deep soil stabilisation* (Bredenberg H, Holm G and Broms B B, eds), Proceedings of International Conference, Stockholm, pp 185–191 (Balkema, Rotterdam)

INGLES O C and METCALF J B (1972)
Soil stabilisation – principles and practice (Butterworths, Sydney)

INSTITUTION OF CIVIL ENGINEERS SITE INVESTIGATION STEERING GROUP (1993)
Site investigation in construction (Thomas Telford, London)

ISHIHARA K and FURUKAWAZONO K (1999)
Performances of storage tanks during the 1995 Kobe earthquake, in *Earthquake geotechnical engineering* (Seco e Pinto P S, ed), Proceedings of 2nd International Conference, Lisbon, vol 3, pp 795–808 (Balkema, Rotterdam)

JAMIOLKOWSKI M, PRESTI D C F LO and FROIO F (1998)
Design parameters of granular soils from *in situ* tests, in *Geotechnical hazards*
(Maric B, Lisac Z and Szavits-Nossan A, eds) Proceedings of 11th Danube-European
Conference on Soil Mechanics and Geotechnical Engineering, Porec, Croatia,
pp 65–94 (Balkema, Rotterdam)

JAPANESE GEOTECHNICAL SOCIETY (ed) (1998)
Remedial measures against soil liquefaction: from investigation and design to
implementation (Balkema, Rotterdam)

JOHNSON D, NICHOLLS R and THOMSON G R (1983)
An evaluation of ground improvement at Belawan Port, North Sumatra, in
Improvement of ground (Rathmayer H G and Saari K H O, eds), Proceedings of 8th
European Conference on Soil Mechanics and Foundation Engineering, Helsinki, vol 1,
pp 45–52 (Balkema, Rotterdam)

JONES D B, MADDISON J D AND BEASLEY D H (1995)
Long-term performance of clay and peat treated by surcharge, *Geotechnical*
Engineering, Proceedings of Institution of Civil Engineers, vol 113, no 1, January,
pp 31–37

KATAOKA K, GOTO T, OGINO T, SCHIMIZU K and NAKANO K (1992)
Bearing capacity of a soil-cement column in soft ground, in *Soil improvement, Current*
Japanese Materials Research, vol 9, pp 149–166 (Elsevier, London)

KIM S-I and KIM D-S (1997) SASW method for the evaluation of ground
densification by dynamic compaction, in *Ground improvement geosystems:*
densification and reinforcement (Davies M C R and Schlosser F, eds) Proceedings of
3rd International Conference, London, pp 141–147 (Thomas Telford, London)

KJELLMAN W (1948)
Accelerating consolidation of fine-grained soils by means of cardboard wicks, in
Proceedings of 2nd International Conference on Soil Mechanics and Foundation
Engineering, Rotterdam, vol 2, pp 302–305

KJELLMAN W (1952)
Consolidation of clay soil by means of atmospheric pressure, in *Proceedings of*
Conference on Soil Stabilisation, MIT, Boston, Massuchussetts, pp 258–263

KOEHORST B A N and VAN DEN BERG A (1999)
The performance of stabilised soil columns in two Dutch test sites, in *Dry Mix*
Methods for Deep Soil Stabilisation, Proceedings of International Conference,
Stockholm, pp 239–244 (Balkema, Rotterdam)

LADE P V and YAMAMURO J A (eds) (1998)
Physics and mechanics of soil liquefaction, Proceedings of International Workshop,
Baltimore, Maryland, September 1998 (Balkema, Rotterdam, 1999)

LAMBE T W and WHITMAN R V (1979)
Soil mechanics, SI version (Wiley, New York)

LAMPERT D and WOODLEY D R (eds) (1991)
Site selection and investigation – a practical handbook, Gower, Aldershot, Hants

LEACH B A and GOODGER H K (1991)
Building on derelict land, Special Publication 78 (CIRIA, London)

LUNNE T, ROBERTSON P K and POWELL J J M (1997)
Cone penetration testing in geotechnical practice (Blackie, London)

MADDISON J D, JONES D B, BELL A L and JENNER C G (1996)
Design and performance of an embankment supported using low strength geogrids and vibro concrete columns, in *Geosynthetics: applications, design and construction* (De Groot M B, Den Hoedt G and Termaat R J, eds), Proceedings of 1st European Geosynthetics Conference, Maastricht, pp 325–332 (Balkema, Rotterdam)

MAIR R J and WOOD D M (1987)
Pressuremeter testing: methods and interpretation, CIRIA B003 (CIRIA, London)

MARCHETTI S (1980)
In situ tests by flat dilatometer *ASCE Journal of Geotechnical Engineering*, vol 106, no GT3, pp 299–321

MATTHEWS M C, HOPE V S and CLAYTON C R I (1996)
The use of surface waves in the determination of ground stiffness profiles, *Geotechnical Engineering*, Proceedings of Institution of Civil Engineers, vol 119, no 2, April, pp 84–95

MATTHEWS M C, CLAYTON C R I and OWN Y (2000)
The use of field geophysical techniques to determine geotechnical stiffness parameters, *Geotechnical Engineering*, Proceedings of Institution of Civil Engineers, vol 143, no 1, January, pp 31–42

MAYNE P W, JONES J S and DUMAS J C (1984)
Ground response to dynamic compaction, *ASCE Journal of Geotechnical Engineering*, vol 110, no 6, June, pp 757–774

MEIGH A C (1987)
Cone penetration testing: methods and interpretation CIRIA B002 (CIRIA, London)

MENARD L and BROISE Y (1975)
Theoretical and practical aspects of dynamic consolidation, *Geotechnique*, vol 25, no 1, March, pp 3–18

MEYERHOF G G (1953)
Some recent foundation research and its application to design, *The Structural Engineer*, vol 31, no 6, pp 151–167

MITCHELL J K (1981)
Soil improvement – state-of-the-art report, in *Proceedings of 10th International Conference on Soil Mechanics and Foundation Engineering*, Stockholm, vol 4, pp 509–565

MITCHELL J K (1986)
Ground improvement evaluation by *in situ* tests, in *Use of in situ tests in geotechnical engineering* (Clemence S P, ed), Proceedings of ASCE Speciality Conference, Virginia, pp 221–236, Geotechnical Special Publication no 6 (ASCE, New York)

MITCHELL J K and BRANDON T L (1998)
Analysis and use of CPT in earthquake and environmental engineering, in
Geotechnical site characterisation (Robertson P K and Mayne P W eds), Proceedings
of 1st International Conference, Atlanta, vol 1, pp 69–97 (Balkema, Rotterdam)

MOSELEY M P (ed) (1993)
Ground Improvement (Blackie, Glasgow)

MOSELEY M P and PRIEBE H J (1993)
Vibro techniques, in *Ground improvement* (Moseley M P ed), pp 1–19 (Blackie,
Glasgow)

MUNFAKH G A, SARKAR S K and CASTELLI R J (1983)
Performance of a test embankment founded on stone columns, in *Piling and Ground
Treatment*, Proceedings of International Conference, London, March 1983, pp 259–265
(Thomas Telford, London, 1984)

MUSSON R M W and WINTER P W (1996) *Seismic hazard of the UK*, Report for
Department of Trade and Industry (AEA Technology, Warrington)

MUSSON R M W and WINTER P W (1997) Seismic hazard maps for the UK, *Natural
Hazards*, vol 14, pp 141–154

NICHOLSON D P, TSE C-M and PENNY C (1999) *The observational method in
ground engineering: principles and applications*, CIRIA Report 185 (CIRIA, London)

O'BRIEN A S (1997)
Vibrocompaction of loose estuarine sands, in *Ground improvement geosystems:
densification and reinforcement* (Davies M C R and Schlosser F, eds), Proceedings of
3rd International Conference, London, pp 121–126 (Thomas Telford, London)

O'BRIEN A S and ANDERSON G (1999)
Design and construction of the UK's first polystyrene embankment for railway use, in
Railway Engineering '99, Proceedings of 2nd International Conference, London
(Engineering Technics Press, London)

OVE ARUP AND PARTNERS (1993)
Preliminary study of UK seismic hazard and risk, Report for Department of the
Environment

PARK C L, JEONG H J, PARK J B, LEE S W, KIM Y S and KIM S J (1997)
A case study of vacuum pre-loading with vertical drains, in *Ground improvement
geosystems: densification and reinforcement* (Davies M C R and Schlosser F, eds),
Proceedings of 3rd International Conference, London, pp 68–74 (Thomas Telford,
London)

PARSONS A W (1992)
*Compaction of soils and granular materials: a review of research performed at the
Transport Research Laboratory* (HMSO, London)

PATEL D (1995)
Opening speaker at British Geotechnical Society Meeting on "Building on fill" held at
Institution of Civil Engineers, 12th April 1995, Meeting report by K S Watts, *Ground
Engineering*, July/August, pp 32–35

POLSHIN D E and TOKAR R A (1957)
Maximum allowable non-uniform settlement of structures, in *Proceedings of 4th International Conference on Soil Mechanics and Foundation Engineering*, London, vol 1, pp 402–405

POWELL J J M and UGLOW I M (1988)
Marchetti dilatometer testing in UK soils, in *Penetration Testing 1988* (J de Ruiter ed), *Proceedings of 1st International Symposium*, ISOPT-1, Orlando, Florida, vol 1, pp 555–562 (Balkema, Rotterdam)

PREENE M, ROBERTS T O L, POWRIE W and DYER M R (2000)
Groundwater control: design and practice, CIRIA C515 (CIRIA, London)

PRIEBE H J (1995)
The design of vibro replacement, *Ground Engineering*, vol 28, December, pp 31–37

RADHAKRISHNAN R, NG FOOK WAH, RAJENDRA A S and PUI S K (1983)
Land reclamation for Singapore Changi airport, in *Reclamation 83*, Proceedings of International Land Reclamation Conference, Grays, Essex, pp 141–150, Industrial Seminars, Tunbridge Wells

RAISON C A (1996)
Contribution to discussion, in *The observational method in geotechnical engineering*, pp 175–179 (Thomas Telford, London)

RAJU V R (1997) The behaviour of very soft cohesive soils improved by vibro replacement, in *Ground improvement geosystems: densification and reinforcement* (Davies M C R and Schlosser F, eds), Proceedings of 3rd International Conference, London, pp 253–259 (Thomas Telford, London)

RAMASWAMY S D and YONG K Y (1983) Evaluation of densification of sandfill, in *Improvement of ground* (Rathmayer H G and Saari K O, eds), Proceedings of 8th European Conference on Soil Mechanics and Foundation Engineering, Helsinki, vol 1, pp 69–72 (Balkema, Rotterdam)

RAWLINGS C G, HELLAWELL E E and KILKENNY W M (2000)
Grouting for ground engineering, CIRIA C514 (CIRIA, London)

SEED H B (1987)
Design problems in soil liquefaction, *ASCE Journal of Geotechnical Engineering*, vol 113, no 8, pp 827–845

SHANG J Q, TANG M and MIAO Z (1998)
Vacuum pre-loading consolidation of reclaimed land: a case study, *Canadian Geotechnical Journal*, vol 35, no 5, October, pp 740–749

SIMPSON B and DRISCOLL R (1998)
Eurocode 7 – a commentary (Construction Research Communications, London)

SKEMPTON A W (1986)
Standard penetration test procedures and the effects in sands of overburden pressures, relative density, particle size, ageing and overconsolidation, *Geotechnique*, vol 36, no 3, pp 425–447

SKEMPTON A W and MACDONALD D H (1956)
The allowable settlement of buildings, *Proceedings of Institution of Civil Engineers*, part 3, vol 5, pp 727–768

SKINNER H D, WATTS K S and CHARLES J A (1997)
Building on colliery spoil: some geotechnical considerations, *Ground Engineering*, vol 30, no 5, June, pp 35–40

SKINNER H D and CHARLES J A (1999) Problems associated with building on a variable depth of fill, *Ground Engineering*, vol 32, no 7, July, pp 32–35

SKINNER H D, CHARLES J A and WATTS K S (1999)
Ground deformations and stress redistribution due to a reduction in volume of zones of soil at depth, *Geotechnique*, vol 49, no 1, February, pp 111–126

SLOCOMBE B C (1993)
Dynamic compaction, In *Ground Improvement* (M P Moseley, ed), pp 20–39 (Blackie, Glasgow)

SLOCOMBE B C, BELL A L and BAEZ J I (2000)
The densification of granular soils using vibro methods, *Geotechnique*, Vol 50, No 6, December, pp 715–725

SMITH M R and COLLIS L (eds) (1993)
Aggregates: sand, gravel and crushed rock aggregates for construction purposes (second edition), Engineering Geology Special Publication no 9 (Geological Society, London)

SMYTH-OSBOURNE K R and MIZON D H (1984) Settlement of a factory on opencast backfill in *Ground movements and structures* (Geddes J D, ed), Proceedings of 3rd International Conference, Cardiff, July, 1984, pp 463–479 (Pentech Press, London, 1985)

SOWERS G F, WILLIAMS R C and WALLACE T S (1965) Compressibility of broken rock and the settlement of rockfills, in *Proceedings of 6th International Conference on Soil Mechanics and Foundation Engineering*, Montreal, vol 2, pp 561–565

SOWERS G B and SOWERS G F (1970) *Introductory Soil Mechanics and Foundations* (3rd edition) (Macmillan, New York)

SOYEZ B, MAGNAN J P and DELFAUT A (1983)
Loading tests on a clayey hydraulic fill stabilized by lime treated soil columns, in *Improvement of ground*, Proceedings of 8th European Conference on Soil Mechanics and Foundation Engineering, Helsinki, vol 2, pp 951–954 (Balkema, Rotterdam)

STAMATOPOULOS A C and KOTZIAS P C (1985)
Soil improvement by preloading (Wiley, New York)

STARK T D and OLSON S M (1995)
Liquefaction resistance using CPT and field case histories, *ASCE Journal of Geotechnical Engineering*, vol 121, no 12, pp 856–869

TERASHI M AND TANAKA H (1981)
Ground improvement by deep mixing method, in *Proceedings of 10th International Conference on Soil Mechanics and Foundation Engineering*, Stockholm, vol 3, pp 777–780 (Balkema, Rotterdam)

TERZAGHI K (1956)
Discussion on The allowable settlement of buildings, *Proceedings of Institution of Civil Engineers*, part 3, vol 5, pp 775–777

TERZAGHI K and PECK R B (1948)
Soil mechanics in engineering practice (Wiley, New York)

THOMPSON R P (1998)
The value of timely hazard identification, in *The value of geotechnics in construction*, Proceedings of Seminar held at Institution of Civil Engineers, pp 3–11 (CRC, London)

TOMLINSON M J (1995)
Foundation design and construction (6th edition) (Longman Scientific and Technical, Harlow, Essex)

TOMLINSON M J and WILSON D M (1973)
Pre-loading of foundations by surcharge on filled ground, *Geotechnique*, vol 23, no 1, March, pp 117–120

TONKS D M, HILLIER R P and BEEDEN H J (1998)
Lime/cement treatment to improve marginal materials, in *The value of geotechnics in construction*, Proceedings of Seminar held at Institution of Civil Engineers, pp 157–167 (CRC, London)

TOTH P S (1993)
In situ soil mixing, in *Ground improvement* (M P Moseley, ed), pp 193–204 (Blackie, Glasgow)

TRENTER N A and CHARLES J A (1996)
A model specification for engineered fills for building purposes, *Geotechnical Engineering*, Proceedings of Institution of Civil Engineers, vol 119, no 4, October, pp 219–230

TURNER M J (1997)
Integrity testing in piling practice, CIRIA R144 (CIRIA, London)

VAN IMPE W F (1989)
Soil improvement techniques and their evolution (Balkema, Rotterdam)

VAN SANTVOORT G P T M (1994)
Geotextiles and geomembranes in civil engineering (revised edition) (Balkema, Rotterdam)

WATTS K S, CHARLES J A and BUTCHER A P (1989)
Ground improvement for low-rise housing using vibro at a site in Manchester, *Municipal Engineer*, vol 6, no 3, pp 145–157

WATTS K S and CHARLES J A (1991)
The use, testing and performance of vibrated stone columns in the United Kingdom, in *Deep foundation improvement, design, construction and testing*, Proceedings of ASTM Symposium, Las Vegas, January 1990, pp 212–233 (STP 1089, ASTM, Philadelphia, 1991)

WATTS K S, SAADI A, WOODS L A and JOHNSON D (1992)
Preliminary report on a field trial to assess the design and performance of vibro ground treatment with reinforced strip foundations, in *Proceedings of 2nd International Conference on Polluted and Marginal Land*, pp 215–221 (Brunel University, London)

WATTS K S and CHARLES J A (1993)
Initial assessment of new rapid ground compactor, in *Engineered Fills* (Clarke B G, Jones C J F P and Moffat A I B, eds), Proceedings of International Conference, Newcastle-upon-Tyne, pp 399–412 (Thomas Telford, London)

WATTS K S and CHARLES J A (1999)
Settlement characteristics of landfill wastes, *Geotechnical Engineering*, Proceedings of Institution of Civil Engineers, vol 137, October, pp 225–233

WATTS K S and SERRIDGE C J (2000)
A trial of vibro bottom-feed stone column treatment in soft clay soil, in *Grouting, Soil Improvement, Geosystems including Reinforcement* (Rathmayer H, ed), Proceedings of 4th International Conference on Ground Improvement Geosystems, Helsinki, pp 549–556 (Building Information Ltd, Helsinki)

WATTS K S, JOHNSON D, WOOD L A and SAADI A (2000)
An instrumented trial of vibro ground treatment supporting strip foundations in a variable fill, *Geotechnique*, Vol 50, No 6, December, pp 699–708

WEST J M (1975)
The role of ground improvement in foundation engineering, *Geotechnique*, vol 25, no 1, pp 71–78

WEST G and CARDER D R (1997)
Review of lime piles and lime-stabilised soil columns, TRL Report 305 (Transport Research Laboratory, Crowthorne)

WILDE P M and CROOK J M (1991)
The monitoring of ground movements and their effects on surface structures – a series of case histories, in *Ground Movements and Structures* (Geddes J D, ed) Proceedings of 4th International Conference, Cardiff, July 1991, pp 182–189 (Pentech Press, London, 1992)

WOLSKI W (1996)
Staged construction, in *Embankments on organic soils* (Hartlen J and Wolski W, eds), pp 293–354 (Elsevier, Amsterdam)

WOOD L A, JOHNSON D, WATTS K S and SAADI A (1996)
Performance of strip footings on fill materials reinforced by stone columns, *The Structural Engineer*, vol 74, no 16, August, pp 265–271

XANTHAKOS P P, ABRAMSON L W AND BRUCE D A (1994)
Ground control and improvement (Wiley, New York)

ZDANKIEWICZ J and WAHAB R M (1999)
Stone column and vibro-compaction of liquefiable deposits at a bridge approach, in *Earthquake geotechnical engineering* (Seco e Pinto P S, ed), Proceedings of 2nd International Conference, Lisbon, vol 2, pp 605–610 (Balkema, Rotterdam)